The Best American Science
and Nature Writing 2014

The Best American Science and Nature Writing™ 2014

Edited and with an Introduction
by Deborah Blum

Tim Folger, Series Editor

A Mariner Original

HOUGHTON MIFFLIN HARCOURT

BOSTON • NEW YORK 2014

ISSN 1530-1508
ISBN 978-0-544-00342-2

Printed in the United States of America
DOC 10 9 8 7 6 5 4 3 2 1

Contents

Contents

Foreword

IN THE SUMMER of 2008, I spent a cool, foggy day hiking in the coastal hills of Wales with Sir John Houghton, one of the world's leading climate scientists. As we tramped through wet, sheep-nibbled grass, the landscape seemed to dissolve into the heavy mists that closed around us, bounding our horizons. Every now and then, through breaks in the fog, we glimpsed vistas of green, treeless hills sloping toward the Irish Sea. Then the view would fade to gray again, stranding us in a world drained of color.

Over the course of that damp day we talked about another sort of shrinking horizon, one composed of fast-vanishing opportunities to prevent irreversible and catastrophic climate change. At the time of our hike it still seemed possible that the world's nations might manage to limit greenhouse gas emissions to levels that would not prove utterly disruptive. Most scientists believe the planet's average global temperature must rise no more than 2 degrees Celsius above preindustrial levels if we are to minimize the hazards a hotter world will bring: melting ice sheets, rising seas, extreme weather, extinctions, crop failures, severe droughts.

Houghton served for nearly fifteen years as one of the lead scientists on the Intergovernmental Panel on Climate Change, overseeing the release of the world's most authoritative reports on the subject. I asked him how much time we had to turn things around. "We have seven years," he said. To confine global warming to 2 degrees Celsius, our emissions of greenhouse gases would have to peak in 2015 and decline steadily thereafter.

We'll surely miss that deadline, and in the meantime we're setting the worst sorts of records. In May 2013 the amount of carbon dioxide in the atmosphere reached 400 parts per million, a level unmatched since 3 million years ago, when sea levels may have been 60 feet higher than today. That dreary benchmark should have made headlines, trumping every other story. Instead, some of our leading news organizations have cut back on environmental coverage. Why does the most important subject of our time receive such underwhelming attention? I sometimes fantasize that daily reports of carbon dioxide concentrations will displace the latest figures from the Dow Jones or that dancers at a Super Bowl halftime spectacle will enact how sea-level rise will force the Miami Dolphins to seek a new home before the century ends.

The world we knew is literally disappearing. Hurricane Sandy chopped more than 30 feet off New Jersey's beaches, redrawing the state's coastline overnight. Such storms will become more frequent and more devastating in the decades ahead. What will it take to wake us up? Perhaps some of the remarkable stories in this collection will help. Few writers have done more than Elizabeth Kolbert to alert us to the enormous scale and peril of the threat we've unleashed on ourselves. This volume contains her magnificent two-part article from *The New Yorker,* "The Lost World," which should be required reading for politicians in every nation. (One hesitates to call them leaders.) Kolbert shows how we are now witnessing ecological changes in mere years that once took place over geologic time scales.

Call me biased, but I'm convinced that you will find in these pages the most important journalism of our time, the stories that will last. Whether it's Nicholas Carr's account of the corrosive effect of technology on our brains in "The Great Forgetting" or Corey S. Powell's "The Madness of the Planets," a lively and lyrical narrative of the dangers lurking in our rough-and-tumble solar system, the articles here are the sort that will change the way you look at the world. They also make for delightful and often moving reading. I know I will never forget the planetary scientist Sarah Stewart Johnson's "O-Rings," with its unexpected links between Antarctic exploration and . . . well, I won't spoil it for you. I was astonished to learn from Johnson that "O-Rings" is her first published essay. I'm certain it won't be her last.

All these stories were chosen by Deborah Blum, one of the best science journalists in the business. I've been a fan ever since I first read her Pulitzer Prize–winning series of articles *The Monkey Wars* more than twenty years ago. I feel privileged to share these pages with her. I try to read widely when searching for articles for this collection, but Blum opened my eyes to a number of wonderful new publications, which I hope will find a wider audience. Now it's time for all the members of that audience to settle into their favorite chairs or curl up in bed and discover twenty-six stories that matter.

I hope too that readers, writers, and editors will nominate their favorite articles for next year's anthology at http://timfolger.net/forums. The criteria for submissions and deadlines, and the address to which entries should be sent, can be found in the "news and announcements" forum on my web site. Once again this year I'm offering an incentive to enlist readers to scour the nation in search of good science and nature writing: send me an article that I haven't found, and if the article makes it into the anthology, I'll mail you a free copy of next year's edition. What do you think, Deborah? Can I get you to sign those copies? I also encourage readers to use the forums to leave feedback about the collection and to discuss all things scientific. The best way for publications to guarantee that their articles are considered for inclusion in the anthology is to place me on their subscription list, using the address posted in the "news and announcements" section of the forums.

I'd like to thank Deborah Blum for putting together such a compelling collection this year. If you'd like to keep up with her writing, I encourage you to follow "Poison Pen," her *New York Times* blog, and "Elemental," her blog for *Wired* magazine. Once again this year I'm indebted to Nina Barnett and her colleagues at Houghton Mifflin Harcourt, who make this collection possible. And I'm most grateful of all for my wise and beauteous *khaleesi*, Anne Nolan. Where would I be without you?

Tim Folger

Introduction

LET'S BEGIN with a summer afternoon in Nevada some twenty-five years ago, hot and quiet, except for the faint, sly rustle of the wind in the desert sand.

Back then I was obsessively following a story about flawed designs of nuclear weapons. In the midst of my pursuit, the military scheduled an underground test of one of the warheads in our stockpile of nuclear armaments. The test was to be conducted at the Nevada Test Site (now called the Nevada National Security Site), a well-guarded stretch of rocky desert northwest of Las Vegas.

I wanted to be there. Somehow the bomb testers didn't think that was a good idea. But they did agree to a guided tour before the detonation, a chance to walk the underground tunnel that led to ground zero. My photographer and I had to get FBI clearance to go on the tour. The waiting period led me to neurotically wonder whether my teenage antiwar-protesting days would become an issue. I had skipped high school to attend some remarkably peaceful marches at the University of Georgia. Now suddenly I was a conspiracy theorist. "They used to fly airplanes over the protests to take photographs of the protesters, you know," I told my husband, a former U.S. Army journalist, in a hissing way. He rolled his eyes. And rolled them again when the clearance sailed through.

The tunnel, as I remember, was quiet, a narrow cave of rough, dark rock, lit by the blue glow of fluorescence. Cables and boxy machinery had been strung down its length to the point where the warhead would wait. They'd dug it deep under one of the mountainous outcroppings that line the Nevada desert, and standing there, you could imagine how the place would shake when the detonation occurred. The rock floor beneath your feet would shatter and the mountain would shimmy, dancing on the desert. But still I was more interested in what lay aboveground, the relics of the far more primitive tests from earlier days.

In the 1950s and early 1960s, the United States tested warheads aboveground. A hundred mushroom clouds blossomed, mostly here in Nevada, before we moved everything underground. Tourists in Las Vegas would gather to watch these demonstrations, applauding as the atomic storm flashed upward, lighting the sky, darkening it. They say the clouds were visible a hundred miles away. We might shake our heads today over such nuclear tourism, but people weren't so aware back then of the invisible drift of radioactive particles and the dangers they carried. That awareness would come later, along with a rise in radiation-induced illnesses across a swath of Nevada. In the moment, though, there was mostly a celebratory sense of dominance, pride that we could harness this wild blaze of power.

Meanwhile, scientists continued to calculate, study the extent of that power. In conjunction with those showy explosions, experimenters at the test site built houses and banks, bridges and stores, set at varying differences from ground zero so they could measure the range and power of the bomb's blast furnace.

On the day of our visit, the landscape of the old "bomb towns" was part of the driving tour. Through the car windows we could see the metal supports of a bridge twisted into black tumbleweed, brick houses with doors and windows blown out by the wind. We parked some distance away, not too far from a crater carved in the desert floor by an old impact. The test site is 680 square miles of mostly wild terrain, and it was moonscape quiet as we stepped out of the car.

"What's that yellow tape for?" I asked, pointing to some rough rectangles blocked off by what looked like crime-scene tape.

"Those are radioactive hot spots," our guide answered.

Only a few minutes later the wind came up in a faint howl, dust wrapped around us like a ground fog, and through this brown-gray mist the guide's voice reassured us: "Don't worry. We'll check you out for radiation levels before you leave."

*

So there you have the slightly alarmed science writer standing in a dirt cloud. Let's leave her there for the moment. There's a reason those blowing particles still drift through my memory. Who doesn't remember those edgy moments when you really wish you'd been standing somewhere else? But this moment also serves as a different kind of reminder. That everything—including a haze of wind-blown dust—is something more, holds a story worth telling.

"To see the world in a grain of sand," wrote the nineteenth-century British poet William Blake in a poem with the lovely title "Auguries of Innocence." I do not mean to claim anything so grand for the point I am making here. But perhaps I can make a case for a small metaphor. This is the natural world reshaped by our activities—a stretch of desert still radioactive because over this dusty terrain we developed, tested, and demonstrated our theories of atomic destruction. Here in this blowing dust is a story of science with all its human determination and innovation—and its occasional hubris. If told right, it's a great story, one that reminds us of all the unexpected complications that often come with scientific advances. In the best science stories, or so I believe, one often sees the arc of the choices we've made as a species to build and create —and occasionally to destroy.

In the stories we tell, the ones that really do justice to the scientific process, we show our readers that arc—the curving, complicated line that links discovery and development, choice and consequence. It sounds like such a simple thing. But it's when we connect the dots that we can connect with our readers in a richer sense. We remind them of the role that research and its results play in their own lives. And we remind them that scientific exploration of the world around us can reveal a portrait of connected lives on a tiny and far too fragile planet. An essay in this book, "Trapline," by David Treuer, an Ojibwe writer, makes that point eloquently. There's a moment in the story when he's holding down an unconscious fox: "I felt the quickness of his breath as I knelt on him with one knee. With one hand on his head and the other on his chest, I felt his heart and the life in it." And even unconscious, the fox's fear knocks against Treuer's hand. "Who knew that a heart could beat that fast?" And he goes on to recognize that in such a place, his own heart might do the same.

The science writers whom I admire bring such connections to singing, stinging life. They remind us that this is ever a human exercise—that scientists, like the rest of us, are just people trying to understand the world around us. They never forget that, as with any human enterprise, mistakes are made and opportunities lost. But they also remind us that at its best, the community of science is one that works to correct its errors, that seeks to make the world right and even better. In today's best science writing you find all of that—the stumbles and the hopes, the unexpected ideas and unexpected beauty. And you find it across an almost limitless spectrum of hard questions, fascinating ideas, sometimes unexpected answers. Even within the limits of this anthology, I can promise you stories that range from the shimmer of deep space to the wayward nature of a wild sheep.

It's been both a pleasure and a humbling experience to read and select the science stories for this anthology. I want to thank *The Best American Science and Nature Writing*'s series editor, Tim Folger, for doing the searching and sifting that produced an amazing selection of articles for me to read. Tim has a wonderful eye for a story that matters. The time I spent reading through the selections brightened some ice-gray winter days in my home state of Wisconsin—which will tell you how good they were, because this last winter was very gray and *very* icy.

So it was difficult to winnow down to the twenty-six stories in this book. They are an eclectic mix, on a variety of subjects—yes, ranging from sheep to stars—in publications including well-established magazines, such as *National Geographic,* and newly created digital ones, such as *Nautilus* and *Medium.* For a longtime science writer like myself, it's reassuring to be reminded that such work continues at traditional outlets, and it's equally exciting to see the rise of innovative new outlets. All of them produced stories that challenged me to think in new ways about science and how it changes the world.

I believe the best science writing does exactly that—encourages us to think and rethink, puts us on a path toward change at a time when we need to take such a path, as the landscape itself changes around us. Some of these stories brought a different perspective to a long-standing issue like climate change (think about the appearance of new hybrid species); some startled me with an idea I hadn't considered before, such as the power of television soap operas to influence birthrates in developing countries. But always they made the world more interesting; they made the journeys of the scientists themselves more real. Sometimes, of course, the science writers are themselves scientists. So you'll find here the famed biologist E. O. Wilson, who is a two-time winner of the Pulitzer Prize for his nonfiction books. In his tale of a national park in Mozambique struggling to recover from the collateral damage inflicted by war, Wilson reminds us to look beyond the poster-child species when we consider the natural world. "People yearn to see large wild animals and I am no exception," he says. "But wildlife also includes the little things that run the world"—a reminder that I think is exactly right, and a phrase that I love. You'll also find the novelist Barbara Kingsolver, once a science writer herself. (In the 1980s she had a science-writing job at the University of Arizona.) Science and nature still weave their way through many of Kingsolver's novels, and in the small, lyrical essay here, she uses knitting as a metaphor for the shifting seasons and textures of life around us: "the particular green-silver of leaves overturned by an oncoming storm. An alkaline desert's russet bronze, a mustard of Appalachian spring, some bright spectral intangible you find you long to possess."

There was pleasure too in discovering writers that I didn't know so well. For instance, those sheep I mentioned. In "Twelve Ways of Viewing Alaska's Wild, White Sheep," Bill Sherwonit weaves together a story of sometimes luminous rock-climbing animals, the hunters who kill them, and the scientists who study them, including the sheep's rather wonderfully shifty response to hunting season. It's a fascinating portrait of a wild animal usually ignored in favor of the region's more dramatic species, like grizzlies—and a portrait of us as well. Both science and self-reflection shine as well in former astronomer Pippa Goldschmidt's essay, "What Our Telescopes Couldn't See," and in Sarah Stewart Johnson's lovely "O-Rings," a story that travels from the frozen Antarctic to the catastrophic launch of the space shuttle *Challenger.* As she makes that journey, Johnson threads through it a meditation on life, death, physics, and the reasons we make daredevil choices.

In many ways, these and the other pieces gathered here are stories of choices and of consequences. You'll see that in Seth Mnookin's "The Return of Measles." Measles is a formidable virus. Its transmission rate can be 90 percent, and it's durable enough

to survive outside the body for some hours, meaning that everything touched by a carrier—even tables and chairs—can harbor
the infection. Mnookin counts, in meticulous detail, what the ill-
informed antivaccine movement may cost us in hospitalizations,
deaths, and the consumption of those increasingly scarce public
health dollars. You'll see choice and consequence in Kate Sheppard's "Under Water," another meticulous accounting, this time
of building and rebuilding on floodplains. Sheppard's story, for
Mother Jones, is about the politics and money involved in coastal
building. It simmers with the frustration of scientists who keep
warning about the costs of building the same homes over and
over again—homes that are increasingly likely to be washed away.
One study found that homes rebuilt more than once because of
flood damage accounted for some 40 percent of National Flood
Insurance Program payouts. "No surprise then," Sheppard writes,
"that the federal insurance program is now $25 billion in the
hole."

You'll see choice and consequence echoing too in Amy Harmon's tale of a Florida orange grower's desperate quest to save
his orchards from a tree-crippling bacterial disease. The disease,
called citrus greening, has relentlessly crept from continent to
continent. Scientists have not found a single citrus species that is
resistant to infection. The grower profiled by Harmon realizes that
the only answer may be in genetic modification. As he pursues
that goal, to save his trees, he finds himself increasingly entangled
in the angry political debate that currently swirls around the issue of genetic engineering. Harmon balances the fraught politics
with the science in the most rational way, debunking some of the
common myths about genetic modification and letting the reader
consider the choice that must eventually be made—and its consequences.

In that same regard, I want to mention Nicholas Carr's "The
Great Forgetting." Carr explores the way that our reliance on such
helpful technologies as GPS mapping makes us less reliant on our
own abilities and knowledge. As we depend on the device rather
than ourselves, that dependence changes us as well. I've long wondered about the consequences of GPS mapping because I use it
every time I visit my son in Chicago and am required to drive the
city's tangle of streets. Although I've made numerous visits, and

driven many miles there, I still have no sense of the city's geography, and without looking at a real map can't tell you exactly where my son's neighborhood is. I might argue that I don't need to, since I use GPS, but sometimes I am uneasily aware that I used to be able to visualize the cities I traveled. Carr's point, though, is far more urgent than mine. He investigates airplane crashes in which the pilot's overdependence on autopilot contributes to fatal errors. "We're forgetting how to fly," one veteran pilot says. Carr also tells us how some traditional cultures, such as the ice-hunting Inuit of Canada, have become so reliant on GPS that they've lost the ability to find their way on their own. They are starting to lose, as one observer puts it, their feel for the land.

Sometimes it seems that we're always slightly behind in this game of choice and consequence. We move forward with all the excited cheer of a new technology or biological insight, and then we realize that we've made our move without fully considering nature's countermove. Perhaps nothing illustrates that better than our overuse of antibiotics and the resulting tide of antibiotic resistance, told in chilling though beautiful detail in Maryn McKenna's look toward a "post-antibiotic" future. The best of popular science writing does many things well—illuminating complicated research, forgotten corners of the earth, the worlds that lie beyond our own—in ways that make the universe itself more real to us. But I've come to believe that it's the ability to see a discovery as a decision, to follow it from start to sometimes troubling, sometimes triumphant finish, that is one of the most important things we science writers bring to the story of science.

It is, of course, that very story of choice and consequence that eddies in my memory of radioactive dust, stirred into the air on that long-ago summer day.

So I'll return you now to that afternoon, finally allowing the science writer, her photographer, and their Nevada test-site guide to leave that drift of bomb-town dust behind. Aside from wishing to be elsewhere, of course, there was nothing to do but wait for the wind to give it up. We brushed ourselves off—no doubt making sure that the dust was all over our hands—and continued the tour.

The day was sunny, our guide was reassuringly nonchalant, and we headed down a strip of narrow roadway to our next destination, which turned out to be the test site's nuclear waste storage area. Metal barrels stamped with radiation symbols were stacked around us. I still have a slightly crackly photograph that my photographer insisted on taking. It shows a woman in her early thirties, wearing a baseball hat, T-shirt, and blue jeans, holding a traditional reporter's notebook in one hand, looking into the camera with a slight smile. Behind her is nothing but a wall of radioactive waste barrels, piled so high that there's no sky visible in the image.

"What were you thinking?" I say to her sometimes. But I know the answer. She was thinking what an incredible place this was, what a reminder of our atomic legacy. We've never really known what to do with the radioactive wastes of nuclear bombs except box them up and hide them away. Fifteen years ago the government created the Waste Isolation Pilot Plant and placed such nuclear detritus from its test facilities in a deep salt mine near Carlsbad, New Mexico. In late February of this year, a leak was reported there; thirteen workers tested positive for radiation exposure. The exposure was low, but it's a reminder that no problem ever remains fully buried. The lesson I took away from the Nevada site, or one of them, is that sometimes the most important thing the writer does is just tell the story, bring the choice and the consequence out of hiding, so that it's not invisible, so that we, as a society, don't forget.

And as you can tell, I haven't forgotten that windblown day. I haven't forgotten because when we went through the radiation detectors on our way out, the alarms started clanging like a fire truck as my photographer passed through. I could feel my eyes go wide, and I know his were as he looked back at me. They sent him through again, and this time the alarms were silent. "Just a little glitch," the guards manning the exit assured us, and this time I could feel my eyes narrowing. But "Just a glitch," I repeated to the photographer as we sped back toward the casino-hotel in Las Vegas where we were staying. The desert whipped by the windows in a blur of russet and gray-green. I had my foot down on the gas because, well, I was in a hurry to get back.

"You know," I said casually as I hopped out of the car, "I think I'll just go take a shower." As I recall, I said it to his back because he was already on his way. When we met for dinner, we were both shiny from soap and water. Over the chicken, though, he decided that he was still so stressed out he might just try relaxing in the glitter of the casino. That night, playing blackjack, he lost every dollar we'd brought to cover our expenses.

But that, as they say, is another story.

DEBORAH BLUM

*The Best American Science
and Nature Writing 2014*

KATHERINE BAGLEY

Mixed Up

FROM *Audubon*

"I LOVE GETTING huge boxes of blood," says the genetic orni-
thologist Rachel Vallender as she pulls open a drawer full of small
plastic vials in her laboratory at the Canadian Museum of Nature
in Ottawa, where she's a visiting scientist. Each tube, carefully labeled and organized, holds a blood sample from a single warbler.
Whether the bird is actually a hybrid is the question Vallender
seeks to answer.

Hybrids of golden-winged and blue-winged warblers are increasingly popping up across the Northeast and into Canada. The
physical differences between the mixed progeny and their pure
counterparts can be subtle. A bird might, for instance, have the
distinctive yellow patches on its wings, the golden head, and the
jet-black collar of a golden-winged warbler but with the yellowish
belly of a blue-winged warbler. So individual scientists and conservation groups, including Audubon North Carolina and Bird
Studies Canada, are gathering samples from across eastern North
America and sending them to Vallender, who analyzes mitochon-
drial DNA in the blood to determine the birds' genetic history.
She examines the shipments she receives in free moments—on
nights, weekends, and vacation days from her full-time job with
Environment Canada, a government agency. The research is revealing how prevalent this intermingling of genes is and helping
bring to light some of the potential dangers it poses.

Records of blue-wingeds spreading into golden-winged territory, hybridizing with them, and gradually replacing them extend

back to the early twentieth century. Such mixing isn't unusual in the avian world: nearly 10 percent of all bird species are known to occasionally interbreed. But the genetic work of Vallender, who has been studying warbler hybridization for more than a decade, backs up the observations of birders and scientists who, during the same time period, have reported growing numbers of hybrids while conducting population surveys. She's found that in many places across the United States and Canada, hybrids now make up as much as 30 percent of golden-winged warbler populations. "This isn't just some sporadic event anymore," she says.

This shift, says Vallender, correlates with the onslaught of climate change. Biologists have long known that habitat loss is a major factor driving blue-winged warblers to expand their range. The bird's preferred scrubland habitat is disappearing as abandoned farmland reverts to forest. Warming temperatures might be adding additional pressures, causing blue-wingeds to move north in search of cooler climes and into habitat already occupied by golden-wingeds.

For reasons unknown, the golden-winged warblers seem to suffer more from the interaction. While blue-winged populations are experiencing declines, golden-winged populations are plummeting, and scientists are wary of the species' chances for long-term survival. "If [this decline] continues at the rate it has been going, we could see drastic reductions in their populations or, worst-case scenario, extinctions," says Vallender. "We need to do this research now."

What's happening to the two warblers isn't unique. Polar bears and grizzly bears are mating, as are different species of everything from butterflies to sharks.

In some instances, it's clear that climate change is playing a role. More than 1,700 animal species across the globe have shifted their ranges northward and upward in elevation, searching for colder temperatures and following as the plants and other animals they rely on shift as well. Ice sheets and other physical barriers that once kept species apart are disappearing. All of these changes are expected to accelerate as we spew ever more greenhouse gases into the atmosphere, driving up the earth's temperature.

Climate-driven intermixing is raising challenging conservation issues. Should hybrid offspring be protected if one parent species is threatened or endangered? Ecologically, does it matter if the world loses purebred species to hybridization? Is it best to get involved or to let nature take its human-altered course, creating new species and eliminating others? These are the questions experts are just beginning to ponder, even as the planet continues to warm.

In 2006 an American big-game hunter from Idaho shot and killed the first documented wild polar–grizzly bear hybrid, a mostly white male covered in patches of brown fur, with long grizzly-like claws, a humped back, and eyes ringed by black skin. Four years later a second-generation "pizzly" or "grolar" was shot. After hearing reports of the bears, Brendan Kelly, then an Alaska-based biologist with the National Oceanic and Atmospheric Administration, started to wonder which other species might be interbreeding as a result of a changing Arctic landscape.

Snow and sea ice hit record lows in 2012, and the Arctic has warmed more than 3.6 degrees Fahrenheit since the mid-1960s, more than twice the global average.

To gauge what kinds of effects these shifts were having on Arctic animals, Kelly teamed up with the biologist David Tallmon at the University of Alaska and the conservation geneticist Andrew Whiteley at the University of Massachusetts Amherst. The trio coauthored a seminal report for the journal *Nature* in 2010 that chronicled the hybridization that wildlife managers and First Nations communities had been seeing in the Arctic, including the mixing of beluga whales and narwhals, bowhead and right whales, Dall's and harbor porpoises, hooded and harp seals, spotted and harbor seals, and North Atlantic minke and North Pacific minke whales, in addition to polar and grizzly bears. They also outlined the devastating effects the new genetic exchanges could have on biodiversity, such as parent species being driven to extinction or creating hybrids unable to survive in the environments they are born into.

The scientific community at large quickly recognized that the genetic mixing wasn't limited to animals in the rapidly changing Arctic. Today they're finding it all over the place, in owls, petrels, squirrels, big cats, and wild canines.

Between 2007 and 2009, researchers from several Australian universities caught fifty-seven hybrid blacktip sharks while doing routine marine surveys off the northeast coast of Australia. Genetic tests confirmed that they were crossbreeds of Australian and common blacktips. The result of several generations of interbreeding, they were found south of the tropical areas where Australian blacktips typically live.

Elsewhere, scientists are discovering that hybridizing species are exchanging behavioral and physiological traits, not just physical ones. Mark Scriber is an entomologist and professor emeritus at Michigan State University who studies swallowtail butterflies. In 1999 he began noticing hybrids in northern ranges that could and were eating plants previously tolerated only by southern swallowtail species. He also discovered hybrids in the north whose emergence had been delayed by four or five weeks, so that they arrived too late to mate with the previous generation of butterflies and too early to mate with the next. They could mate only with each other, essentially creating a new species.

These sorts of interactions are, in their purest form, a kind of evolution, points out Kelly. For millennia, wildlife was forced together and pushed apart as climate, ecosystems, and landscapes changed. During these periods of upheaval, genes flowed between animals, creating new species and driving others to extinction. But genetic mixing that frequently takes centuries now takes only decades or even years, because modern climate change is altering the earth so quickly and drastically.

Regardless of the cause, Jim Mallet, an evolutionary biologist at Harvard University who has studied hybridization in European and South American butterflies, argues that we should let nature take its course. And while he isn't completely alone in his thinking, most other scientists interviewed for this story were divided over whether to take action or let the interactions play out unimpeded. "My feeling is that hybridization is natural," Mallet says. "It is the result of a mating decision by an individual, and different individuals have different desires and interests. You don't want to label a mating decision as unnatural when it's found in the wild."

Still, recent human-driven hybridization could have catastrophic results for species. "The climate warming that we have induced is closer to a meteor strike [for species] than to the gradual evolution of green plants," says Kelly, who is now the assistant director of polar science for the White House's Office of Science and Technology Policy. "We're forcing change to happen so quickly that it is more likely to promote extinctions than provide adaptive responses."

Unnaturally speeding up the hybridization process can significantly affect biodiversity and the animals themselves. Pairings in which one parent species is threatened usually hasten its decline, though scientists aren't certain why one set of genes wins out over the other, as Vallender has seen with blue-winged warblers surviving while golden-winged warblers die off.

Across the continent, Eric Forsman, a biologist with the U.S. Forest Service, has watched closely while spotted owls, a threatened species in the United States, lose their tenuous foothold in the Pacific Northwest, in part because of interbreeding with newly arrived barred owls. Barred owls have expanded their range from their native Midwestern homeland to the Pacific Coast, likely due to wildfires and climate change. "One hypothesis is that because of warming temperatures, the forests of northern Canada expanded," scattering woodlands across the Great Plains and creating a migration corridor, Forsman says. "That may have allowed barred owls to expand westward across what was once a physical barrier and into spotted owl territory." What spurs the two species to interbreed isn't well understood, says Forsman. It may be that since there are fewer of their own species to mate with, spotted owls pair up with the first owl they see—and that's likely to be one of the many barred owls that have moved into the area. Most barred owls, meanwhile, continue to mate with their own species. The end result is fewer spotted owls.

The more genetically similar two species are—in terms of chromosome numbers or reproductive proteins—the easier it is for them to reproduce. Dogs and cats, for example, or lions and lambs are just too different genetically to produce offspring. But even when interspecies pairing is successful, the hybrid offspring face a series of unique challenges. Just as a mule—a cross between a male donkey and a female horse—is usually sterile, many hybrids cannot reproduce and are therefore genetic dead ends, says Mallet. Others inherit traits from their parents that render them ill equipped to thrive or even survive. Polar–grizzly bear hybrids bred in captivity, for instance, can't swim as well as genetically pure polar bears, which could pose grave risks in an ecosystem where ice sheets—the frozen platforms from which they hunt seals—are smaller and farther apart.

One of the biggest debates is about whether hybrids should be eligible for legal protection, particularly if one or both parent species are threatened or endangered. Currently, the Endangered Species Act doesn't address hybrids. The same goes for the International Union for Conservation of Nature's Red List of Threatened Species. (Hybrids may often be unknowingly protected because they can be difficult to distinguish from their safeguarded parents.) The U.S. Fish and Wildlife Service drafted a hybrid policy in 1996 but ultimately decided not to approve it, says J. B. Ruhl, an environmental lawyer and expert in climate change and the Endangered Species Act at Vanderbilt Law School. The agency instead adopted a policy of dealing with these animals on a case-by-case basis. Neither a Fish and Wildlife Service press officer nor several conservation lawyers could name any hybrids currently protected by the agency.

The problem, Ruhl says, is that the Fish and Wildlife Service's current policy addresses individuals, while the real issue is populations.

Scientists and conservation experts are split as to whether these legal policies should be changed to deal with the growing number of hybrids. Some see no value in keeping hybrids around at all. Stuart Pimm, a species extinction expert at Duke University, says that wiping out hybrids is the best way to protect threatened species—though doing so would be tricky, he admits. "An unfortunate aspect of all this is that hybridization is a major cause of species endangerment and disappearance," he says. "This is not one of those circumstances where the choices are easy ones, but these hybrids are a threat to many valued species." Hybrids have value, too, argues Richard Kock, a conservationist and member of the IUCN's Species Survival Commission. "We should see [hybrids] as holding genes, some of which represent original species and therefore are of value," he says. "With modern genetic understanding, breeding back to an original genotype is not impossible. So they have a place in conservation."

Ultimately, how we deal with hybrids will be decided among lawmakers and wildlife managers, in courtrooms and at international meetings. "It becomes a value question," says Kelly. "Do you like having a white bear that specializes in hunting seals in the ice? That's what's in peril."

NICHOLAS CARR

The Great Forgetting

FROM *The Atlantic*

ON THE EVENING of February 12, 2009, a Continental Connection commuter flight made its way through blustery weather between Newark, New Jersey, and Buffalo, New York. As is typical of commercial flights today, the pilots didn't have all that much to do during the hour-long trip. The captain, Marvin Renslow, manned the controls briefly during takeoff, guiding the Bombardier Q400 turboprop into the air, then switched on the autopilot and let the software do the flying. He and his copilot, Rebecca Shaw, chatted —about their families, their careers, the personalities of air-traffic controllers—as the plane cruised uneventfully along its northwesterly route at 16,000 feet. The Q400 was well into its approach to the Buffalo airport, its landing gear down, its wing flaps out, when the pilot's control yoke began to shudder noisily, a signal that the plane was losing lift and risked going into an aerodynamic stall. The autopilot disconnected, and the captain took over the controls. He reacted quickly, but he did precisely the wrong thing: he jerked back on the yoke, lifting the plane's nose and reducing its airspeed, instead of pushing the yoke forward to gain velocity. Rather than preventing a stall, Renslow's action caused one. The plane spun out of control, then plummeted. "We're down," the captain said, just before the Q400 slammed into a house in a Buffalo suburb.

The crash, which killed all forty-nine people on board as well as one person on the ground, should never have happened. A National Transportation Safety Board investigation concluded that the cause of the accident was pilot error. The captain's response to the stall warning, the investigators reported, "should have been automatic, but his improper flight control inputs were inconsistent with his training" and instead revealed "startle and confusion." An executive from the company that operated the flight, the regional carrier Colgan Air, admitted that the pilots seemed to lack "situational awareness" as the emergency unfolded.

The Buffalo crash was not an isolated incident. An eerily similar disaster, with far more casualties, occurred a few months later. On the night of May 31, an Air France Airbus A330 took off from Rio de Janeiro, bound for Paris. The jumbo jet ran into a storm over the Atlantic about three hours after takeoff. Its air-speed sensors, coated with ice, began giving faulty readings, causing the autopilot to disengage. Bewildered, the pilot flying the plane, Pierre-Cedric Bonin, yanked back on the stick. The plane rose and a stall warning sounded, but he continued to pull back heedlessly. As the plane climbed sharply, it lost velocity. The airspeed sensors began working again, providing the crew with accurate numbers. Yet Bonin continued to slow the plane. The jet stalled and began to fall. If he had simply let go of the control, the A330 would likely have righted itself. But he didn't. The plane dropped 35,000 feet in three minutes before hitting the ocean. All 228 passengers and crew members died.

The first automatic pilot, dubbed a "metal airman" in a 1930 *Popular Science* article, consisted of two gyroscopes, one mounted horizontally, the other vertically, that were connected to a plane's controls and powered by a wind-driven generator behind the propeller. The horizontal gyroscope kept the wings level, while the vertical one did the steering. Modern autopilot systems bear little resemblance to that rudimentary device. Controlled by onboard computers running immensely complex software, they gather information from electronic sensors and continuously adjust a plane's attitude, speed, and bearings. Pilots today work inside what they call "glass cockpits." The old analog dials and gauges are mostly gone. They've been replaced by banks of digital displays. Automation has become so sophisticated that on a typical passenger flight, a human pilot holds the controls for a grand total of just three minutes. What pilots spend a lot of time doing is monitoring screens and keying in data. They've become, it's not much of an exaggeration to say, computer operators.

And that, many aviation and automation experts have con-

cluded, is a problem. Overuse of automation erodes pilots' expertise and dulls their reflexes, leading to what Jan Noyes, an ergo-
nomics expert at Britain's University of Bristol, terms "a de-skilling
of the crew." No one doubts that autopilot has contributed to improvements in flight safety over the years. It reduces pilot fatigue
and provides advance warnings of problems, and it can keep a
plane airborne should the crew become disabled. But the steady
overall decline in plane crashes masks the recent arrival of "a spec-
tacularly new type of accident," says Raja Parasuraman, a psychology professor at George Mason University and a leading authority
on automation. When an autopilot system fails, too many pilots,
thrust abruptly into what has become a rare role, make mistakes.
Rory Kay, a veteran United Airlines captain who has served as the
top safety official of the Air Line Pilots Association, put the problem bluntly in a 2011 interview with the Associated Press: "We're
forgetting how to fly." The Federal Aviation Administration has
become so concerned that in January 2013 it issued a "safety alert"
to airlines, urging them to get their pilots to do more manual flying. An overreliance on automation, the agency warned, could put
planes and passengers at risk.

The experience of airlines should give us pause. It reveals that
automation, for all its benefits, can take a toll on the performance
and talents of those who rely on it. The implications go well beyond safety. Because automation alters how we act, how we learn,
and what we know, it has an ethical dimension. The choices we
make, or fail to make, about which tasks we hand off to machines
shape our lives and the place we make for ourselves in the world.
That has always been true, but in recent years, as the locus of
labor-saving technology has shifted from machinery to software,
automation has become ever more pervasive, even as its workings
have become more hidden from us. Seeking convenience, speed,
and efficiency, we rush to offload work to computers without reflecting on what we might be sacrificing as a result.

Doctors use computers to make diagnoses and to perform surgery. Wall Street bankers use them to assemble and trade financial
instruments. Architects use them to design buildings. Attorneys
use them in document discovery. And it's not only professional
work that's being computerized. Thanks to smartphones and
other small, affordable computers, we depend on software to carry
out many of our everyday routines. We launch apps to aid us in

shopping, cooking, socializing, even raising our kids. We follow turn-by-turn GPS instructions. We seek advice from recommendation engines on what to watch, read, and listen to. We call on Google, or Siri, to answer our questions and solve our problems. More and more, at work and at leisure, we're living our lives inside glass cockpits.

A hundred years ago, the British mathematician and philosopher Alfred North Whitehead wrote, "Civilization advances by extending the number of important operations which we can perform without thinking about them." It's hard to imagine a more confident expression of faith in automation. Implicit in Whitehead's words is a belief in a hierarchy of human activities: Every time we offload a job to a tool or a machine, we free ourselves to climb to a higher pursuit, one requiring greater dexterity, deeper intelligence, or a broader perspective. We may lose something with each upward step, but what we gain is, in the long run, far greater.

History provides plenty of evidence to support Whitehead. We humans have been handing off chores, both physical and mental, to tools since the invention of the lever, the wheel, and the counting bead. But Whitehead's observation should not be mistaken for a universal truth. He was writing when automation tended to be limited to distinct, well-defined, and repetitive tasks—weaving fabric with a steam loom, adding numbers with a mechanical calculator. Automation is different now. Computers can be programmed to perform complex activities in which a succession of tightly coordinated tasks is carried out through an evaluation of many variables. Many software programs take on intellectual work—observing and sensing, analyzing and judging, even making decisions—that until recently was considered the preserve of humans. That may leave the person operating the computer to play the role of a high-tech clerk—entering data, monitoring outputs, and watching for failures. Rather than opening new frontiers of thought and action, software ends up narrowing our focus. We trade subtle, specialized talents for more routine, less distinctive ones.

Most of us want to believe that automation frees us to spend our time on higher pursuits but doesn't otherwise alter the way we behave or think. That view is a fallacy—an expression of what scholars of automation call the "substitution myth." A labor-saving device doesn't just provide a substitute for some isolated component of a job or other activity. It alters the character of the entire task, including the roles, attitudes, and skills of the people taking part. As Parasuraman and a colleague explained in a 2010 journal article, "Automation does not simply supplant human activity but rather changes it, often in ways unintended and unanticipated by the designers of automation."

Psychologists have found that when we work with computers, we often fall victim to two cognitive ailments—complacency and bias—that can undercut our performance and lead to mistakes. Automation complacency occurs when a computer lulls us into a false sense of security. Confident that the machine will work flawlessly and handle any problem that crops up, we allow our attention to drift. We become disengaged from our work, and our awareness of what's going on around us fades. Automation bias occurs when we place too much faith in the accuracy of the information coming through our monitors. Our trust in the software becomes so strong that we ignore or discount other information sources, including our own eyes and ears. When a computer provides incorrect or insufficient data, we remain oblivious to the error.

Examples of complacency and bias have been well documented in high-risk situations—on flight decks and battlefields, in factory control rooms—but recent studies suggest that the problems can bedevil anyone working with a computer. Many radiologists today use analytical software to highlight suspicious areas on mammograms. Usually the highlights aid in the discovery of disease. But they can also have the opposite effect. Biased by the software's suggestions, radiologists may give cursory attention to the areas of an image that haven't been highlighted, sometimes overlooking an early-stage tumor. Most of us have experienced complacency when at a computer. In using e-mail or word-processing software, we become less proficient proofreaders when we know that a spell checker is at work.

The way computers can weaken awareness and attentiveness points to a deeper problem. Automation turns us from actors into observers. Instead of manipulating the yoke, we watch the screen. That shift may make our lives easier, but it can also inhibit the development of expertise. Since the late 1970s, psychologists have been documenting a phenomenon called the "generation effect." It was first observed in studies of vocabulary, which revealed that people remember words much better when they actively call them to mind—when they generate them—than when they simply read them. The effect, it has since become clear, influences learning in many different circumstances. When you engage actively in a task, you set off intricate mental processes that allow you to retain more knowledge. You learn more and remember more. When you repeat the same task over a long period, your brain constructs specialized neural circuits dedicated to the activity. It assembles a rich store of information and organizes that knowledge in a way that allows you to tap into it instantaneously. Whether it's Serena Williams on a tennis court or Magnus Carlsen at a chessboard, an expert can spot patterns, evaluate signals, and react to changing circumstances with speed and precision that can seem uncanny. What looks like instinct is hard-won skill, skill that requires exactly the kind of struggle that modern software seeks to alleviate.

In 2005, Christof van Nimwegen, a cognitive psychologist in the Netherlands, began an investigation into software's effects on the development of know-how. He recruited two sets of people to play a computer game based on a classic logic puzzle called Missionaries and Cannibals. To complete the puzzle, a player has to transport five missionaries and five cannibals (or, in van Nimwegen's version, five yellow balls and five blue ones) across a river, using a boat that can accommodate no more than three passengers at a time. The tricky part is that cannibals must never outnumber missionaries, either in the boat or on the riverbanks. One of van Nimwegen's groups worked on the puzzle using software that provided step-by-step guidance, highlighting which moves were permissible and which weren't. The other group used a rudimentary program that offered no assistance.

As you might expect, the people using the helpful software made quicker progress at the outset. They could simply follow the prompts rather than having to pause before each move to remember the rules and figure out how they applied to the new situation. But as the test proceeded, those using the rudimentary software gained the upper hand. They developed a clearer conceptual understanding of the task, plotted better strategies, and made fewer mistakes. Eight months later, van Nimwegen had the same people work through the puzzle again. Those who had earlier used the rudimentary software finished the game almost twice as quickly as their counterparts. Enjoying the benefits of the generation effect, they displayed better "imprinting of knowledge."

What van Nimwegen observed in his laboratory—that when we automate an activity, we hamper our ability to translate information into knowledge—is also being documented in the real world. In many businesses, managers and other professionals have come to depend on decision-support systems to analyze information and suggest courses of action. Accountants, for example, use the systems in corporate audits. The applications speed the work, but some signs suggest that as the software becomes more capable, the accountants become less so. One recent study, conducted by Australian researchers, examined the effects of systems used by three international accounting firms. Two of the firms employed highly advanced software that, based on an accountant's answers to basic questions about a client, recommended a set of relevant business risks to be included in the client's audit file. The third firm used simpler software that required an accountant to assess a list of possible risks and manually select the pertinent ones. The researchers gave accountants from each firm a test measuring their expertise. Those from the firm with the less helpful software displayed a significantly stronger understanding of different forms of risk than did those from the other two firms.

What's most astonishing, and unsettling, about computer automation is that it's still in its early stages. Experts used to assume that there were limits to the ability of programmers to automate complicated tasks, particularly those involving sensory perception, pattern recognition, and conceptual knowledge. They pointed to the example of driving a car, which requires not only the instantaneous interpretation of a welter of visual signals but also the ability to adapt seamlessly to unanticipated situations. "Executing a left turn across oncoming traffic," two prominent economists wrote in 2004, "involves so many factors that it is hard to imagine the set of rules that can replicate a driver's behavior." Just six years later, in October 2010, Google announced that it had built a fleet of seven "self-driving cars," which had already logged more than 140,000 miles on roads in California and Nevada.

Driverless cars provide a preview of how robots will be able to navigate and perform work in the physical world, taking over activities requiring environmental awareness, coordinated motion, and fluid decision making. Equally rapid progress is being made in automating cerebral tasks. Just a few years ago, the idea of a computer competing on a game show like *Jeopardy* would have seemed laughable, but in a celebrated match in 2011, the IBM supercomputer Watson trounced *Jeopardy*'s all-time champion, Ken Jennings. Watson doesn't think the way people think; it has no understanding of what it's doing or saying. Its advantage lies in the extraordinary speed of modern computer processors.

In *Race Against the Machine,* a 2011 e-book on the economic implications of computerization, the MIT researchers Erik Brynjolfsson and Andrew McAfee argue that Google's driverless car and IBM's Watson are examples of a new wave of automation that, drawing on the "exponential growth" in computer power, will change the nature of work in virtually every job and profession. Today, they write, "computers improve so quickly that their capabilities pass from the realm of science fiction into the everyday world not over the course of a human lifetime, or even within the span of a professional's career, but instead in just a few years."

Who needs humans anyway? That question, in one rhetorical form or another, comes up frequently in discussions of automation. If computers' abilities are expanding so quickly and if people, by comparison, seem slow, clumsy, and error-prone, why not build immaculately self-contained systems that perform flawlessly without any human oversight or intervention? Why not take the human factor out of the equation? The technology theorist Kevin Kelly, commenting on the link between automation and pilot error, argued that the obvious solution is to develop an entirely autonomous autopilot: "Human pilots should not be flying planes in the long run." The Silicon Valley venture capitalist Vinod Khosla recently suggested that health care will be much improved when medical software—which he has dubbed "Doctor Algorithm"— evolves from assisting primary-care physicians in making diagnoses to replacing the doctors entirely. The cure for imperfect automation is total automation.

That idea is seductive, but no machine is infallible. Sooner or later, even the most advanced technology will break down, misfire, or, in the case of a computerized system, encounter circumstances that its designers never anticipated. As automation technologies become more complex, relying on interdependencies among algorithms, databases, sensors, and mechanical parts, the potential sources of failure multiply. They also become harder to detect. All of the parts may work flawlessly, but a small error in system design can still cause a major accident. And even if a perfect system could be designed, it would still have to operate in an imperfect world.

In a classic 1983 article in the journal *Automatica,* Lisanne Bainbridge, an engineering psychologist at University College London, described a conundrum of computer automation. Because many system designers assume that human operators are "unreliable and inefficient," at least when compared with a computer, they strive to give the operators as small a role as possible. People end up functioning as mere monitors, passive watchers of screens. That's a job that humans, with our notoriously wandering minds, are especially bad at. Research on vigilance, dating back to studies of radar operators during World War II, shows that people have trouble maintaining their attention on a stable display of information for more than half an hour. "This means," Bainbridge observed, "that it is humanly impossible to carry out the basic function of monitoring for unlikely abnormalities." And because a person's skills "deteriorate when they are not used," even an experienced operator will eventually begin to act like an inexperienced one if restricted to just watching. The lack of awareness and the degradation of know-how raise the odds that when something goes wrong, the operator will react ineptly. The assumption that the human will be the weakest link in the system becomes self-fulfilling.

Psychologists have discovered some simple ways to temper automation's ill effects. You can program software to shift control back to human operators at frequent but irregular intervals; knowing that they may need to take command at any moment keeps people engaged, promoting situational awareness and learning. You can put limits on the scope of automation, making sure that people working with computers perform challenging tasks rather than merely observing. Giving people more to do helps sustain the generation effect. You can incorporate educational routines into software, requiring users to repeat difficult manual and mental tasks that encourage memory formation and skill building.

Some software writers take such suggestions to heart. In schools, the best instructional programs help students master a subject by encouraging attentiveness, demanding hard work, and reinforcing learned skills through repetition. Their design reflects the latest discoveries about how our brains store memories and weave them into conceptual knowledge and practical know-how. But most software applications don't foster learning and engagement. In fact, they have the opposite effect. That's because taking the steps necessary to promote the development and maintenance of expertise almost always entails a sacrifice of speed and productivity. Learning requires inefficiency. Businesses, which seek to maximize productivity and profit, would rarely accept such a trade-off. Individuals, too, almost always seek efficiency and convenience. We pick the program that lightens our load, not the one that makes us work harder and longer. Abstract concerns about the fate of human talent can't compete with the allure of saving time and money.

The small island of Igloolik, off the coast of the Melville Peninsula in the Nunavut territory of northern Canada, is a bewildering place in the winter. The average temperature hovers at about 20 degrees below zero, thick sheets of sea ice cover the surrounding waters, and the sun is rarely seen. Despite the brutal conditions, Inuit hunters have for some four thousand years ventured out from their homes on the island and traveled across miles of ice and tundra to search for game. The hunters' ability to navigate vast stretches of the barren Arctic terrain, where landmarks are few, snow formations are in constant flux, and trails disappear overnight, has amazed explorers and scientists for centuries. The Inuit's extraordinary way-finding skills are born not of technological prowess—they long eschewed maps and compasses—but of a profound understanding of winds, snowdrift patterns, animal behavior, stars, and tides.

Inuit culture is changing now. The Igloolik hunters have begun to rely on computer-generated maps to get around. Adoption of GPS technology has been particularly strong among younger Inuit, and it's not hard to understand why. The ease and convenience of automated navigation makes the traditional Inuit techniques seem archaic and cumbersome.

But as GPS devices have proliferated on Igloolik, reports of serious accidents during hunts have spread. A hunter who hasn't developed way-finding skills can easily become lost, particularly if his GPS receiver fails. The routes so meticulously plotted on satellite maps can also give hunters tunnel vision, leading them onto thin ice or into other hazards a skilled navigator would avoid. The anthropologist Claudio Aporta, of Carleton University in Ottawa, has been studying Inuit hunters for more than fifteen years. He notes that while satellite navigation offers practical advantages, its adoption has already brought a deterioration in way-finding abilities and, more generally, a weakened feel for the land. An Inuit on a GPS-equipped snowmobile is not so different from a suburban commuter in a GPS-equipped SUV: as he devotes his attention to the instructions coming from the computer, he loses sight of his surroundings. He travels "blindfolded," as Aporta puts it. A unique talent that has distinguished a people for centuries may evaporate in a generation.

Whether it's a pilot on a flight deck, a doctor in an examination room, or an Inuit hunter on an ice floe, knowing demands doing. One of the most remarkable things about us is also one of the easiest to overlook: each time we collide with the real, we deepen our understanding of the world and become more fully a part of it. While we're wrestling with a difficult task, we may be motivated by an anticipation of the ends of our labor, but it's the work itself —the means—that makes us who we are. Computer automation severs the ends from the means. It makes getting what we want easier, but it distances us from the work of knowing. As we transform ourselves into creatures of the screen, we face an existential question: Does our essence still lie in what we know, or are we now content to be defined by what we want? If we don't grapple with that question ourselves, our gadgets will be happy to answer it for us.

DAVID DOBBS

The Social Life of Genes

FROM *Pacific Standard*

A FEW YEARS AGO, Gene Robinson, of Urbana, Illinois, asked some associates in southern Mexico to help him kidnap some one thousand newborns. For their victims they chose bees. Half were European honeybees, *Apis mellifera ligustica,* the sweet-tempered kind most beekeepers raise. The other half were *ligustica*'s genetically close cousins, *Apis mellifera scutellata,* the African strain better known as killer bees. Though the two subspecies are nearly indistinguishable, the latter defend territory far more aggressively. Kick a European honeybee hive, and perhaps a hundred bees will attack you. Kick a killer bee hive, and you may suffer a thousand stings or more. Two thousand will kill you.

Working carefully, Robinson's conspirators—researchers at Mexico's National Center for Research in Animal Physiology, in the high resort town of Ixtapan de la Sal—jiggled loose the lids from two African hives and two European hives, pulled free a few honeycomb racks, plucked off about 250 of the youngest bees from each hive, and painted marks on the bees' tiny backs. Then they switched each set of newborns into the hive of the other subspecies.

Robinson, back in his office at the University of Illinois at Urbana-Champaign's Department of Entomology, did not fret about the bees' safety. He knew that if you move bees to a new colony in their first day, the colony accepts them as its own. Nevertheless, Robinson did expect that the bees would be changed by their adoptive homes: he expected the killer bees to take on the European bees' moderate ways and the European bees to assume the killer bees' more violent temperament. Robinson had discovered this in prior experiments. But he hadn't yet figured out how it happened.

He suspected the answer lay in the bees' genes. He didn't expect the bees' actual DNA to change: random mutations aside, genes generally don't change during an organism's lifetime. Rather, he suspected that the bees' genes would behave differently in their new homes—wildly differently.

This notion was both reasonable and radical. Scientists have known for decades that genes can vary their level of activity, as if controlled by dimmer switches. Most cells in your body contain every one of your 22,000 or so genes. But in any given cell at any given time, only a tiny percentage of those genes are active, sending out chemical messages that affect the activity of the cell. This variable gene activity, called gene expression, is how your body does most of its work.

Sometimes these turns of the dimmer switch correspond to basic biological events, as when you develop tissues in the womb, enter puberty, or stop growing. At other times gene activity cranks up or spins down in response to changes in your environment. Thus certain genes switch on to fight infection or heal your wounds—or, running amok, give you cancer or burn your brain with fever. Changes in gene expression can make you thin, fat, or strikingly different from your supposedly identical twin. When it comes down to it, really, genes don't make you who you are. Gene expression does. And gene expression varies depending on the life you live.

Every biologist accepts this. That was the safe, reasonable part of Robinson's notion. Where he went out on a limb was in questioning the conventional wisdom that environment usually causes fairly *limited* changes in gene expression. It might sharply alter the activity of some genes, as happens in cancer or digestion. But in all but a few special cases, the thinking went, environment generally brightens or dims the activity of only a few genes at a time.

Robinson, however, suspected that environment could spin the dials on "big sectors of genes, right across the genome"—and that an individual's social environment might exert a particularly powerful effect. Who you hung out with and how they behaved, in short, could dramatically affect which of your genes spoke up and which stayed quiet—and thus change who you were.

Robinson was already seeing this in his bees. The winter before, he had asked a new postdoc, Cédric Alaux, to look at the gene-expression patterns of honeybees that had been repeatedly exposed to a pheromone that signals alarm. (Any honeybee that detects a threat emits this pheromone. It happens to smell like bananas. Thus "it's not a good idea," says Alaux, "to eat a banana next to a beehive.")

To a bee, the pheromone makes a social statement: *Friends, you are in danger.* Robinson had long known that bees react to this cry by undergoing behavioral and neural changes: their brains fire up and they literally fly into action. He also knew that repeated alarms make African bees more and more hostile. When Alaux looked at the gene-expression profiles of the bees exposed again and again to the alarm pheromone, he and Robinson saw why: with repeated alarms, hundreds of genes—genes that previous studies had associated with aggression—grew progressively busier. The rise in gene expression neatly matched the rise in the aggressiveness of the bees' response to threats.

Robinson had not expected that. "The pheromone just lit up the gene expression, and it kept leaving it higher." The reason soon became apparent: some of the genes affected were transcription factors—genes that regulate other genes. This created a cascading gene-expression response, with scores of genes responding.

This finding inspired Robinson's kidnapping-and-cross-fostering study. Would moving baby bees to wildly different social environments reshape the curves of their gene-expression responses? Down in Ixtapan, Robinson's collaborators suited up every five to ten days, opened the hives, found about a dozen foster bees in each one, and sucked them up with a special vacuum. The vacuum shot them into a chamber chilled with liquid nitrogen. The intense cold instantly froze the bees' every cell, preserving the state of their gene activity at that moment. At the end of six weeks, when the researchers had collected about 250 bees representing every stage of bee life, the team packed up the frozen bees and shipped them to Illinois.

There Robinson's staff removed the bees' sesame-seed-size brains, ground them up, and ran them through a DNA microarray machine. This identified which genes were busy in a bee's brain at the moment it met the bee-vac. When Robinson sorted his data by group—European bees raised in African hives, for instance, or African bees raised normally among their African kin—he could see how each group's genes reacted to their lives.

Robinson organized the data for each group onto a grid of red and green color-coded squares: each square represented a different gene, and its color represented the group's average rate of gene expression. Red squares represented genes that were especially active in most of the bees in that group; the brighter the red, the more bees in which that gene had been busy. Green squares represented genes that were silent or underactive in most of the group. The printout of each group's results looked like a sort of cubist Christmas card.

When he got the cards, says Robinson, "the results were stunning." For the bees that had been kidnapped, life in a new home had indeed altered the activity of "whole sectors" of genes. When their gene-expression data was viewed on the cards alongside the data for groups of bees raised among their own kin, a mere glance showed the dramatic change. Hundreds of genes had flipped colors. The move between hives didn't just make the bees act differently. It made their genes work differently, and on a broad scale.

What's more, the cards for the adopted bees of both species came to ever more resemble, as they moved through life, the cards of the bees they moved in with. With every passing day their genes acted more like those of their new hive mates (and less like those of their genetic siblings back home). Many of the genes that switched on or off are known to affect behavior; several are associated with aggression. The bees also acted differently. Their dispositions changed to match that of their hive mates. It seemed the genome, without changing its code, could transform an animal into something very like a different subspecies.

These bees didn't just act like different bees. They had pretty much become different bees. To Robinson this spoke of a genome far more fluid—far more socially fluid—than previously conceived.

Robinson soon realized he was not alone in seeing this. At conferences and in the literature, he kept bumping into other researchers who saw gene networks responding fast and widely to social life. David Clayton, a neurobiologist also on the University of Illinois campus, found that if a male zebra finch heard another male zebra finch singing nearby, a particular gene in the bird's forebrain would fire up—and it would do so differently depending on whether the other finch was strange and threatening or familiar and safe.

Others found this same gene, dubbed zenk, ramping up in other species. In each case, the change in zenk's activity corresponded to some change in behavior: a bird might relax in response to a song or become vigilant and tense. Duke researchers, for instance, found that when female zebra finches listened to male zebra finches' songs, the females' zenk gene triggered massive gene-expression changes in their forebrains—a socially sensitive brain area in birds as well as humans. The changes differed depending on whether the song was a mating call or a territorial claim. And perhaps most remarkably, all of these changes happened incredibly fast—within a half hour, sometimes within just five minutes.

Zenk, it appeared, was a so-called immediate early gene, a type of regulatory gene that can cause whole networks of other genes to change activity. These sorts of regulatory gene-expression responses had already been identified in physiological systems such as digestion and immunity. Now they also seemed to drive quick responses to social conditions.

One of the most startling early demonstrations of such a response occurred in 2005 in the lab of the Stanford biologist Russell Fernald. For years Fernald had studied the African cichlid *Astatotilapia burtoni,* a freshwater fish about 2 inches long and dull pewter in color. By 2005 he had shown that among *burtoni,* the top male in any small population lives like some fishy pharaoh, getting far more food, territory, and sex than even the No. 2 male. This No. 1 male cichlid also sports a bigger and brighter body. And there is always only one No. 1.

I wonder, Fernald thought, what would happen if we just removed him?

So one day Fernald turned out the lights over one of his cichlid tanks, scooped out big flashy No. 1, and then, twelve hours later, flipped the lights back on. When the No. 2 cichlid saw that *he* was now No. 1, he responded quickly. He underwent massive surges in gene expression that immediately blinged up his pewter coloring with lurid red and blue streaks and, in a matter of hours, caused him to grow some 20 percent. It was as if Jason Schwartzman, coming to work one day and learning that the big office stud had quit, morphed into Arnold Schwarzenegger by close of business.

These studies, says Greg Wray, an evolutionary biologist at Duke who has focused on gene expression for over a decade, caused quite a stir. "You suddenly realize birds are hearing a song and having massive, widespread changes in gene expression in just fifteen minutes? Something big is going on."

This big something, this startlingly quick gene-expression response to the social world, is a phenomenon we are just beginning to understand. The recent explosion of interest in "epigenetics" —a term literally meaning "around the gene" and referring to anything that changes a gene's effect without changing the actual DNA sequence—has tended to focus on the long game of gene-environment interactions: how famine among expectant mothers in the Netherlands during World War II, for instance, affected gene expression and behavior in their children; or how mother rats, by licking and grooming their pups more or less assiduously, can alter the wrappings around their offspring's DNA in ways that influence how anxious the pups will be for the rest of their lives. The idea that experience can echo in our genes across generations is certainly a powerful one. But to focus only on these narrow, long-reaching effects is to miss much of the action concerning epigenetic influence and gene activity. This fresh work by Robinson, Fernald, Clayton, and others—encompassing studies of multiple organisms, from bees and birds to monkeys and humans—suggests something more exciting: that our social lives can change our gene expression with a rapidity, breadth, and depth previously overlooked.

Why would we have evolved this way? The most probable answer is that an organism that responds quickly to fast-changing social environments will more likely survive them. That organism won't have to wait around, as it were, for better genes to evolve on the species level. Immunologists discovered something similar twenty-five years ago: adapting to new pathogens the old-fashioned way —waiting for natural selection to favor genes that create resistance to specific pathogens—would happen too slowly to counter the rapidly changing pathogen environment. Instead, the immune system uses networks of genes that can respond quickly and flexibly to new threats.

We appear to respond in the same way to our social environment. Faced with an unpredictable, complex, ever-changing population to whom we must respond successfully, our genes behave accordingly—as if a fast, fluid response is a matter of life or death.

About the time Robinson was seeing fast gene-expression changes in bees, in the early 2000s, he and many of his colleagues were taking notice of an up-and-coming UCLA researcher named Steve Cole.

Cole, a Californian then in his early forties, had trained in psychology at UC Santa Barbara and Stanford, then in social psychology, epidemiology, virology, cancer, and genetics at UCLA. Even as an undergrad, Cole had "this astute, fine-grained approach," says Susan Andersen, a professor of psychology now at NYU who was one of his teachers at UC Santa Barbara in the late 1980s. "He thinks about things in very precise detail."

In his postdoctoral work at UCLA, Cole focused on the genetics of immunology and cancer because those fields had pioneered hard-nosed gene-expression research. After that he became one of the earliest researchers to bring the study of whole-genome gene expression to social psychology. The gene's ongoing, real-time response to incoming information, he realized, is where life works many of its changes on us. The idea is both reductive and expansive. We are but cells. At each cell's center, a tight tangle of DNA writes and hands out the cell's marching orders. Between that center and the world stands only a series of membranes.

"Porous membranes," notes Cole.

"We think of our bodies as stable biological structures that live in the world but are fundamentally separate from it. That we are unitary organisms in the world but passing through it. But what we're learning from the molecular processes that actually keep our bodies running is that we're far more fluid than we realize, and the world passes through us."

Cole told me this over dinner. We had met on the UCLA campus and walked south a few blocks, through bright April sun, to an almost empty sushi restaurant. Now, waving his chopsticks over a platter of urchin, squid, and amberjack, he said, "Every day, as our cells die off, we have to replace 1 to 2 percent of our molecular being. We're constantly building and reengineering new cells. And that regeneration is driven by the contingent nature of gene expression.

"This is what a cell is about. A cell," he said, clasping some amberjack, "is a machine for turning experience into biology."

When Cole started his social psychology research in the early 1990s, the microarray technology that spots changes in gene expression was still in its expensive infancy and saw use primarily in immunology and cancer. So he began by using the tools of epidemiology—essentially the study of how people live their lives. Some of his early papers looked at how social experience affected men with HIV. In a 1996 study of eighty gay men, all of whom had been HIV-positive but healthy nine years earlier, Cole and his colleagues found that closeted men succumbed to the virus much more readily.

He then found that HIV-positive men who were lonely also got sicker sooner, regardless of whether they were closeted. Then he showed that closeted men *without* HIV got cancer and various infectious diseases at higher rates than openly gay men did. At about the same time, psychologists at Carnegie Mellon finished a well-controlled study showing that people with richer social ties got fewer common colds.

Something about feeling stressed or alone was gumming up the immune system—sometimes fatally.

"You're besieged by a virus that's going to kill you," says Cole, "but the fact that you're socially stressed and isolated seems to shut down your viral defenses. What's going on there?"

He was determined to find out. But the research methods on hand at the time could take him only so far: "Epidemiology won't exactly lie to you. But it's hard to get it to tell you the whole story." For a while he tried to figure things out at the bench, with pipettes and slides and assays. "I'd take norepinephrine [a key stress hormone] and squirt it on some infected T-cells and watch the virus grow faster. The norepinephrine was knocking down the antiviral response. That's great. Virologists love that. But it's not satisfying as a complete answer, because it doesn't fully explain what's happening in the real world.

"You can make almost anything happen in a test tube. I needed something else. I had set up all this theory. I needed a place to test it."

His next step was to turn to rhesus monkeys, a lab species that allows controlled study. In 2007 he joined John Capitanio, a primatologist at the University of California, Davis, in looking at how social stress affected rhesus monkeys with SIV, or simian immunodeficiency virus, the monkey version of HIV. Capitanio had found that monkeys with SIV fell ill and died faster if they were stressed out by constantly being moved into new groups among strangers — a simian parallel to Cole's 1996 study on lonely gay men.

Capitanio had run a rough immune analysis, which showed that the stressed monkeys mounted weak antiviral responses. Cole offered to look deeper. First he tore apart the lymph nodes — "ground central for infection" — and found that in the socially stressed monkeys, the virus bloomed around the sympathetic nerve trunks, which carry stress signals into the lymph node.

"This was a hint," says Cole: The virus was running amok precisely where the immune response should have been strongest. The stress signals in the nerve trunks, it seemed, were being either muted en route or ignored on arrival. As Cole looked closer, he found it was the latter: the monkeys' bodies were generating the appropriate stress signals, but the immune system didn't seem to be responding to them properly. Why not? He couldn't find out with the tools he had. He was still looking at cells. He needed to look inside them.

Finally Cole got his chance. At UCLA, where he had been made a professor in 2001, he had been working hard to master gene-expression analysis across an entire genome. Microarray machines — the kind Gene Robinson was using on his bees — were getting cheaper. Cole got access to one and put it to work.

Thus commenced what we might call the lonely-people studies.

First, in collaboration with the University of Chicago social psychologist John Cacioppo, Cole mined a questionnaire about social connections that Cacioppo had given to 153 healthy Chicagoans in their fifties and sixties. Cacioppo and Cole identified the eight most socially secure people and the six loneliest and drew blood samples from them. (The socially insecure half-dozen were lonely indeed; they reported having felt distant from others for the previous four years.) Then Cole extracted genetic material from the blood's leukocytes (a key immune-system player) and looked at what their DNA was up to.

He found a broad, weird, strongly patterned gene-expression response that would become mighty familiar over the next few years. Of roughly 22,000 genes in the human genome, the lonely and not-lonely groups showed sharply different gene-expression responses in 209. That meant that about 1 percent of the genome —a considerable portion—was responding differently depending on whether a person felt alone or connected. Printouts of the subjects' gene-expression patterns looked much like Robinson's red-and-green readouts of the changes in his cross-fostered bees: whole sectors of genes looked markedly different in the lonely and the socially secure. And many of these genes played roles in inflammatory immune responses.

Now Cole was getting somewhere.

Normally, a healthy immune system works by deploying what amounts to a leashed attack dog. It detects a pathogen, then sends inflammatory and other responses to destroy the invader while also activating an anti-inflammatory response—the leash—to keep the inflammation in check. The lonely Chicagoans' immune systems, however, suggested an attack dog off leash—even though they weren't sick. Some 78 genes that normally work together to drive inflammation were busier than usual, as if these healthy people were fighting infection. Meanwhile, 131 genes that usually cooperate to control inflammation were underactive. The underactive genes also included key antiviral genes.

This opened a whole new avenue of insight. If social stress reliably created this gene-expression profile, it might explain a lot about why, for instance, the lonely HIV carriers in Cole's earlier studies fell so much faster to the disease.

But this was a study of just fourteen people. Cole needed more.

Over the next several years, he got them. He found similarly unbalanced gene-expression or immune-response profiles in groups including poor children, depressed people with cancer, and people caring for spouses dying of cancer. He topped his efforts off with a study in which social stress levels in young women predicted changes in their gene activity six months later. Cole and his collaborators on that study, psychologists Gregory Miller and Nicolas Rohleder of the University of British Columbia, interviewed 103 healthy Vancouver-area women aged fifteen to nineteen about their social lives, drew blood, and ran gene-expression profiles, and after half a year drew blood and ran profiles again. Some of the women reported at the time of the initial interview that they were having trouble with their love lives, their families, or their friends. Over the next six months, these socially troubled subjects took on the sort of imbalanced gene-expression profile Cole found in his other isolation studies: busy attack dogs and broken leashes. Except here, in a prospective study, he saw the attack dog breaking free of its restraints: social stress changed these young women's gene-expression patterns before his eyes.

In early 2009, Cole sat down to make sense of all this in a review paper that he would publish later that year in *Current Directions in Psychological Science*. Two years later we sat in his spare, rather small office at UCLA and discussed what he'd found. Cole, trimly built but close to 6 feet tall, speaks in a reedy voice that is slightly higher than his frame might lead you to expect. Sometimes, when he's grabbing for a new thought or trying to emphasize a point, it jumps a register. He is often asked to give talks about his work, and it's easy to see why: relaxed but animated, he speaks in such an organized manner that you can almost see the paragraphs form in the air between you. He spends much of his time on the road. Thus the half-unpacked office, he said, gesturing around him. His lab, down the hall, "is essentially one really good lab manager" —Jesusa M. Arevalo, whom he frequently lists on his papers—"and a bunch of robots," the machines that run the assays.

"We typically think of stress as being a risk factor for disease," said Cole. "And it is, somewhat. But if you actually measure stress, using our best available instruments, it can't hold a candle to social isolation. Social isolation is the best-established, most robust social or psychological risk factor for disease out there. Nothing can compete."

This helps explain, for instance, why many people who work in high-stress but rewarding jobs don't seem to suffer ill effects, while others, particularly those isolated and in poverty, wind up accruing lists of stress-related diagnoses—obesity, type 2 diabetes, hypertension, atherosclerosis, heart failure, stroke.

Despite these well-known effects, Cole said he was amazed when he started finding that social connectivity wrought such powerful effects on gene expression.

"Or not that we found it," he corrected, "but that we're seeing it
with such consistency. Science is noisy. I would've bet my eyeteeth
that we'd get a lot of noisy results that are inconsistent from one
realm to another. And at the level of individual genes that's kind
of true—there is some noise there." But the kinds of genes that get
dialed up or down in response to social experience, he said, and
the gene networks and gene-expression cascades that they set off,
"are surprisingly consistent—from monkeys to people, from five-
year-old kids to adults, from Vancouver teenagers to sixty-year-olds
living in Chicago."

Cole's work carries all kinds of implications—some weighty and
practical, some heady and philosophical. It may, for instance, help
explain the health problems that so often haunt the poor. Poverty
savages the body. Hundreds of studies over the past few decades
have tied low income to higher rates of asthma, flu, heart attacks,
cancer, and everything in between. Poverty itself starts to look like
a disease. Yet an empty wallet can't make you sick. And we all know
people who escape poverty's dangers. So what is it about a life of
poverty that makes us ill?

Cole asked essentially this question in a 2008 study he conducted
with Gregory Miller and Edith Chen, another social psychologist
then at the University of British Columbia. The paper appeared
in an odd forum: *Thorax,* a journal about medical problems in the
chest. The researchers gathered and ran gene-expression profiles
on thirty-one kids, ranging from nine to eighteen years old, who
had asthma; sixteen were poor, fifteen well-off. As Cole expected,
the group of well-off kids showed a healthy immune response, with
elevated activity among genes that control pulmonary inflammation. The poorer kids showed busier inflammatory genes, sluggish-
ness in the gene networks that control inflammation, and—in their
health histories—more asthma attacks and other health problems.
Poverty seemed to be mucking up their immune systems.

Cole, Chen, and Miller, however, suspected something else was
at work—something that often came with poverty but was not the
same thing. So along with drawing the kids' blood and gathering
their socioeconomic information, they showed them films of ambiguous or awkward social situations, then asked them how threatening they found them.

The poorer kids perceived more threat; the well-off perceived less. This difference in what psychologists call "cognitive framing" surprised no one. Many prior studies had shown that poverty and poor neighborhoods, understandably, tend to make people more sensitive to threats in ambiguous social situations. Chen in particular had spent years studying this sort of effect.

But in this study, Chen, Cole, and Miller wanted to see if they could tease apart the effect of cognitive framing from the effects of income disparity. It turned out they could, because some of the kids in each income group broke type. A few of the poor kids saw very little menace in the ambiguous situations, and a few well-off kids saw a lot. When the researchers separated those perceptions from the socioeconomic scores and laid them over the gene-expression scores, they found that it was really the kids' framing, not their income levels, that accounted for most of the difference in gene expression. To put it another way: when the researchers controlled for variations in threat perception, poverty's influence almost vanished. The main thing driving screwy immune responses appeared to be not poverty, but whether the child saw the social world as scary.

But where did *that* come from? Did the kids see the world as frightening because they had been taught to, or because they felt alone in facing it? The study design couldn't answer that. But Cole believes isolation plays a key role. This notion gets startling support from a 2004 study of fifty-seven school-age children who were so badly abused that state social workers had removed them from their homes. The study, often just called "the Kaufman study," after its author, the Yale psychiatrist Joan Kaufman, challenges a number of assumptions about what shapes responses to trauma or stress.

The Kaufman study at first looks like a classic investigation into the so-called depression-risk gene—the serotonin transporter gene, or SERT—which comes in both long and short forms. Any single gene's impact on mood or behavior is limited, of course, and these single-gene or "candidate gene" studies must be viewed with that in mind. Yet many studies have found that SERT's short form seems to render many people (and rhesus monkeys) more sensitive to environment; according to those studies, people who carry the short SERT are more likely to become depressed or anxious if faced with stress or trauma.

Kaufman looked first to see whether the kids' mental health

tracked their SERT variants. It did: the kids with the short variant suffered twice as many mental-health problems as those with the long variant. The double whammy of abuse plus short SERT seemed to be too much.

Then Kaufman laid both the kids' depression scores and their SERT variants across the kids' levels of "social support." In this case, Kaufman narrowly defined social support as contact at least monthly with a trusted adult figure outside the home. Extraordinarily, for the kids who had it, this single, modest, closely defined social connection erased about 80 percent of the combined risk of the short SERT variant and the abuse. It came close to inoculating kids against both an established genetic vulnerability and horrid abuse.

Or, to phrase it as Cole might, the lack of a reliable connection harmed the kids almost as much as abuse did. Their isolation wielded enough power to raise the question of what's really most toxic in such situations. Most of the psychiatric literature essentially views bad experiences—extreme stress, abuse, violence—as toxins, and "risk genes" as quasi-immunological weaknesses that let the toxins poison us. And abuse is clearly toxic. Yet if social connection can almost completely protect us against the well-known effects of severe abuse, isn't the isolation almost as toxic as the beatings and neglect?

The Kaufman study also challenges much conventional Western thinking about the state of the individual. To use the language of the study, we sometimes conceive of "social support" as a sort of add-on, something extra that might somehow fortify us. Yet this view assumes that humanity's default state is solitude. It's not. Our default state is connection. We are social creatures, and have been for eons. As Cole's colleague John Cacioppo puts it in his book *Loneliness*, Hobbes had it wrong when he wrote that human life without civilization was "solitary, poor, nasty, brutish, and short." It may be poor, nasty, brutish, and short. But seldom has it been solitary.

Toward the end of the dinner I shared with Cole, after the waiter took away the empty platters and we sat talking over green tea, I asked him if there was anything I should have asked but had not. He'd been talking most of three hours. Some people run dry. Cole does not. He spoke about how we are permeable fluid beings instead of stable unitary isolates; about recursive reconstruction of the self; about an engagement with the world that constantly creates a new you, only you don't know it, because you're not the person you would have been otherwise—you're a one-person experiment that has lost its control.

He wanted to add one more thing: he didn't see any of this as deterministic.

We were obviously moving away from what he could prove at this point, perhaps from what is testable. We were in fact skirting the rabbit hole that is the free-will debate. Yet he wanted to make it clear he does not see us as slaves to either environment or genes.

"You can't change your genes. But if we're even half right about all this, you can change the way your genes behave—which is almost the same thing. By adjusting your environment you can adjust your gene activity. That's what we're doing as we move through life. We're constantly trying to hunt down that sweet spot between too much challenge and too little.

"That's a really important part of this: to an extent that immunologists and psychologists rarely appreciate, we are architects of our own experience. Your subjective experience carries more power than your objective situation. If you feel like you're alone even when you're in a room filled with the people closest to you, you're going to have problems. If you feel like you're well supported even though there's nobody else in sight; if you carry relationships in your head; if you come at the world with a sense that people care about you, that you're valuable, that you're okay; then your body is going to act as if you're okay—even if you're wrong about all that."

Cole was channeling John Milton: "The mind is its own place, and in itself can make a heaven of hell, a hell of heaven."

Of course I did not realize that at the moment. My reaction was more prosaic.

"So environment and experience aren't the same," I offered.

"Exactly. Two people may share the same environment but not the same experience. The experience is what you make of the environment. It appears you and I are both enjoying ourselves here, for instance, and I think we are. But if one of us didn't like being one-on-one at a table for three hours, that person could get quite stressed out. We might have much different experiences. And you can shape all this by how you frame things. You can shape both

your environment and yourself by how you act. It's really an opportunity."

Cole often puts it differently at the end of his talks about this
line of work. "Your experiences today will influence the molecular
composition of your body for the next two to three months," he
tells his audience, "or, perhaps, for the rest of your life. Plan your
day accordingly."

PIPPA GOLDSCHMIDT

What Our Telescopes Couldn't See

FROM *The New York Times*

EDINBURGH — To avoid light pollution and bad weather, professional astronomers have to be prepared to travel long distances to use telescopes on mountaintops far away from towns or cities. Astronomers from Britain, which is not generally well known for its clear skies, are particularly used to traveling. When I was an astronomer in the 1990s, working at the Royal Observatory in Edinburgh and at Imperial College London, I made regular trips to observatories around the world, especially in Chile.

Back home all I could usually see were a few lonely stars fighting against the clouds lit up by a flat dull wash of reflected streetlight. Looking up from the Andes was a fundamentally different experience—the sky was so bright with stars that there were scarcely any dark gaps between them. To make the experience even more striking, in the Southern Hemisphere you can see phenomena like the Magellanic Clouds, the crowded heart of the Milky Way, and the jewel-like constellation of the Southern Cross.

I was studying quasars, some of the most distant known objects in the universe, which are understood to be the bright centers of galaxies containing supermassive black holes. At a superficial glance, quasars don't look particularly interesting. All you can see is a starlike point of light, because the extraordinarily bright center dwarfs the rest of the galaxy. But detailed measurements of quasars reveal their vast distances and the amazingly high speeds of the gases that spin around the invisible black holes.

When I first went to Chile's Atacama Desert in 1990—the year Augusto Pinochet finally stepped down after nearly sixteen years in power—I found that the terrestrial landscape was almost as extraordinary as the objects I studied in the sky. In some parts of the desert it rains only once every few years. During the day there is nothing to see but rocks and sand and the dazzling sunlight reflected from distant telescope domes at the other observatories around the mountains.

At this time, astronomy was on the cusp of changing technologies. I was accustomed to using old-fashioned glass photographic plates—13 inches across and thin enough to be curved into the focal plane of the telescope. The light of stars was caught in the photographic emulsion on the plates, and the resulting complicated shapes of galaxies reminded me of insects preserved in amber.

But this sort of photography was dying out and being superseded by CCDs—electronic devices that created virtual images of the sky, and the precursors of today's digital cameras.

So instead of fitting the plates into the telescope by hand while gazing out at the sky, observing in Chile meant sitting in a control room and operating the telescope remotely. Resulting images were displayed on computer screens. We saw the telescope only at the beginning of the night, when we were setting up, and at the end, when we shut everything down as the sun rose. During the night we didn't even see the sky, because the control room—with its banks of computers that made it look like a low-rent starship *Enterprise*—had no windows.

After a few nights of this work, everything, not just my quasars, started to feel remote. The objects on the screen reminded me of ghosts, with no substance to them. Going outside helped somewhat—the stars looked so large and real I felt I could reach out and grab them—but in contrast the landscape was so empty I might as well have been on the moon. There was nothing to pull me back to earth.

I used to remind myself that I was there to work and nothing more. But I would long for the complexities of life, tedious British rain, and the screech of seagulls, even traffic jams started to have a certain nostalgic charm. As I sat in the control room, I would reread old newspapers, trying to keep faith with the world.

But at that I did not succeed. I know now that the desert wasn't empty, that it had a terrible secret; farther north from the observatory, in an even more remote part of the Andes, was an abandoned prison camp set up during the Pinochet regime, called Chacabuco.

This was a place where political opponents of the regime were sent in the 1970s. Sometimes they died there. As the celebrated 2010 Chilean documentary *Nostalgia for the Light* shows, their relatives are still searching the desert for their remains.

Our telescopes had the power to detect candle flames many miles away, not to mention galaxies billions of light-years away. And yet they never turned toward the camp. They weren't built to do that sort of observation.

Nor were we. The observatory that I visited in Chile is part of the European Southern Observatory. It's one of the largest in the world and was first conceived of not long after World War II, partly as a way of bringing together former enemies and encouraging peaceful scientific endeavors. But our only political contribution was neutrality: we were supposed to keep our heads down and get on with our work. When I arrived, we were advised not to talk about politics with people in Chile.

I left the world of professional astronomy some time ago. In the years since, I have often thought of how astronomy is seen as a benign, unbiased science. Its sole function is to increase our understanding of how the universe works: astronomers receive and record, but they do not experiment or perturb. They are not tainted by any application to, say, energy development or military technologies. Astronomy is, essentially, a passive science.

I remember realizing when I was a student that I could make a measurement of an object in the sky, and how extraordinary that felt to me, as if it were a way of reaching out and connecting with something so far away. But maybe I found distant galaxies easier to understand than the people around me, and I wonder if my work became a substitute for any true connection. I still look to the edge of the universe, but I try to remember always to keep one eye focused here on earth.

AMY HARMON

A Race to Save the Orange
by Altering Its DNA

FROM *The New York Times*

CLEWISTON, FLORIDA — The call Ricke Kress and every other citrus grower in Florida dreaded came while he was driving.

"It's here" was all his grove manager needed to say to force him over to the side of the road.

The disease that sours oranges and leaves them half green, already ravaging citrus crops across the world, had reached the state's storied groves. Kress, the president of Southern Gardens Citrus, in charge of 2.5 million orange trees and a factory that squeezes juice for Tropicana and Florida's Natural, sat in silence for several long moments.

"OK," he said finally on that fall day in 2005, "let's make a plan."

In the years that followed, he and the 8,000 other Florida growers who supply most of the nation's orange juice poured everything they had into fighting the disease they call citrus greening.

To slow the spread of the bacterium that causes the scourge, they chopped down hundreds of thousands of infected trees and sprayed an expanding array of pesticides on the winged insect that carries it. But the contagion could not be contained.

They scoured Central Florida's half-million acres of emerald groves and sent search parties around the world to find a naturally immune tree that could serve as a new progenitor for a crop that has thrived in the state since its arrival, it is said, with Ponce de Leon. But such a tree did not exist.

"In all of cultivated citrus, there is no evidence of immunity," the plant pathologist heading a National Research Council task force on the disease said.

In all of citrus, but perhaps not in all of nature. With a precipitous decline in Florida's harvest predicted within the decade, the only chance left to save it, Kress believed, was one that his industry and others had long avoided for fear of consumer rejection. They would have to alter the orange's DNA—with a gene from a different species.

Oranges are not the only crop that might benefit from genetically engineered resistance to diseases for which standard treatments have proven elusive. And advocates of the technology say it could also help provide food for a fast-growing population on a warming planet by endowing crops with more nutrients or the ability to thrive in drought or to resist pests. Leading scientific organizations have concluded that shuttling DNA between species carries no intrinsic risk to human health or the environment, and that such alterations can be reliably tested.

But the idea of eating plants and animals whose DNA has been manipulated in a laboratory—called genetically modified organisms, or GMOs—still spooks many people. Critics worry that such crops carry risks not yet detected, and they distrust the big agrochemical companies that have produced the few in wide use. And hostility toward the technology, long ingrained in Europe, has deepened recently among Americans as organic food advocates, environmentalists, and others have made opposition to it a pillar of a growing movement for healthier and ethical food choices.

Kress's boss worried about damaging the image of juice long promoted as "100 percent natural."

"Do we really want to do this?" he demanded in a 2008 meeting at the company's headquarters on the northern rim of the Everglades.

Kress, now sixty-one, had no particular predilection for biotechnology. Known for working long hours, he rose through the ranks at fruit and juice companies like Welch's and Seneca Foods. On moving here for the Southern Gardens job, just a few weeks before citrus greening was detected, he had assumed that his biggest headache would be competition from flavored waters or persuading his wife to tolerate Florida's humidity.

But the dwindling harvest that could mean the idling of his juice processing plant would also have consequences beyond any one company's bottom line. Florida is the second-largest producer of orange juice in the world, behind Brazil. Its $9 billion citrus industry contributes 76,000 jobs to the state that hosts the Orange Bowl. Southern Gardens, a subsidiary of U.S. Sugar, was one of the few companies in the industry with the wherewithal to finance the development of a "transgenic" tree, which could take a decade and cost as much as $20 million.

An emerging scientific consensus held that genetic engineering would be required to defeat citrus greening. "People are either going to drink transgenic orange juice or they're going to drink apple juice," one University of Florida scientist told Kress.

And if the presence of a new gene in citrus trees prevented juice from becoming scarcer and more expensive, Kress believed, the American public would embrace it. "The consumer will support us if it's the only way," Kress assured his boss.

His quest to save the orange offers a close look at the daunting process of genetically modifying one well-loved organism—on a deadline. In the past several years, out of public view, he has considered DNA donors from all over the tree of life, including two vegetables, a virus, and, briefly, a pig. A synthetic gene, manufactured in the laboratory, also emerged as a contender.

Trial trees that withstood the disease in his greenhouse later succumbed in the field. Concerns about public perception and potential delays in regulatory scrutiny put a damper on some promising leads. But intent on his mission, Kress shrugged off signs that national campaigns against genetically modified food were gaining traction.

Only in recent months has he begun to face the full magnitude of the gap between what science can achieve and what society might accept.

Millenniums of Intervention

Even in the heyday of frozen concentrate, the popularity of orange juice rested largely on its image as the ultimate natural beverage, fresh squeezed from a primordial fruit. But the reality is that human intervention has modified the orange for millenniums, as it has almost everything people eat.

Before humans were involved, corn was a wild grass, tomatoes were tiny, carrots were only rarely orange, and dairy cows produced little milk. The orange, for its part, might never have existed had human migration not brought together the grapefruit-size pomelo from the tropics and the diminutive mandarin from a temperate zone thousands of years ago in China. And it would not have become the most widely planted fruit tree had human traders not carried it across the globe.

The varieties that have survived, among the many that have since arisen through natural mutation, are the product of human selection, with nearly all of Florida's juice a blend of just two: the Hamlin, whose unremarkable taste and pale color are offset by its prolific yield in the early season, and the dark, flavorful, late-season Valencia.

Because oranges themselves are hybrids, and most seeds are clones of the mother, new varieties cannot easily be produced by crossbreeding—unlike, say, apples, which breeders have remixed into favorites like Fuji and Gala. But the vast majority of oranges in commercial groves are the product of a type of genetic merging that predates the Romans, in which a slender shoot of a favored fruit variety is grafted onto the sturdier roots of other species: lemon, for instance, or sour orange. A seedless midseason orange recently adopted by Florida growers emerged after breeders bombarded a seedy variety with radiation to disrupt its DNA, a technique for accelerating evolution that has yielded new varieties in dozens of crops, including barley and rice.

Its proponents argue that genetic engineering is one in a continuum of ways humans shape food crops, each of which carries risks: even conventional crossbreeding has occasionally produced toxic varieties of some vegetables. Because making a GMO typically involves adding one or a few genes, each containing instructions for a protein whose function is known, they argue, it is more predictable than traditional methods that involve randomly mixing or mutating many genes of unknown function.

But because it also usually involves taking DNA from the species where it evolved and putting it in another to which it may be only distantly related—or turning off genes already present—

critics of the technology say it represents a new and potentially more hazardous degree of tinkering whose risks are not yet fully understood.

If he had had more time, Kress could have waited for the orange to naturally evolve resistance to the bacteria known as *C. liberibacter asiaticus*. That could happen tomorrow. Or it could take years or many decades. Or the orange in Florida could disappear first.

Plunging Ahead

Early discussions among other citrus growers about what kind of disease research they should collectively support did little to reassure Kress about his own genetic engineering project.

"The public will never drink GMO orange juice," one grower said at a contentious 2008 meeting. "It's a waste of our money."

"The public is already eating tons of GMOs," countered Peter McClure, a big grower.

"This isn't like a bag of Doritos," snapped another. "We're talking about a raw product, the essence of orange."

The genetically modified foods Americans have eaten for more than a decade—corn, soybeans, some cottonseed oil, canola oil, and sugar—come mostly as invisible ingredients in processed foods like cereal, salad dressing, and tortilla chips. And the few GMOs sold in produce aisles—a Hawaiian papaya, some squash, a fraction of sweet corn—lack the iconic status of a breakfast drink that, Kress conceded, is "like motherhood" to Americans, who drink more of it per capita than anyone else.

If various polls were to be believed, a third to half of Americans would refuse to eat any transgenic crop. One study's respondents would accept only certain types: two-thirds said they would eat a fruit modified with another plant gene, but few would accept one with DNA from an animal. Fewer still would knowingly eat produce that contained a gene from a virus.

There also appeared to be an abiding belief that a plant would take on the identity of the species from which its new DNA was drawn, like the scientist in the movie *The Fly*, who sprouted insect parts after a DNA-mixing mistake with a housefly.

Asked if tomatoes containing a gene from a fish would "taste fishy" in a question on a 2004 poll conducted by the Food Policy Institute at Rutgers University, referring to one company's efforts to forge a frost-resistant tomato with a gene from the winter flounder, fewer than half correctly answered "no." A fear that the genetic engineering of food would throw the ecosystem out of whack showed in the surveys too.

Kress's researchers, in turn, liked to point out that the very reason genetic engineering works is that all living things share a basic biochemistry: if a gene from a cold-water fish can help a tomato resist frost, it is because DNA is a universal code that tomato cells know how to read. Even the most distantly related species—say, humans and bacteria—share many genes whose functions have remained constant across billions of years of evolution.

"It's not where a gene comes from that matters," one researcher said. "It's what it does."

Kress set the surveys aside.

He took encouragement from other attempts to genetically modify foods that were in the works. There was even another fruit, the "Arctic apple," whose genes for browning were switched off to reduce waste and allow the fruit to be more readily sold sliced.

"The public is going to be more informed about GMOs by the time we're ready," Kress told his research director, Michael P. Irey, as they lined up the five scientists whom Southern Gardens would underwrite. And to the scientists, growers, and juice processors at a meeting convened by Minute Maid in Miami in early 2010, he insisted that just finding a gene that worked had to be his company's priority.

The foes were formidable. *C. liberibacter*, the bacterium that kills citrus trees by choking off their flow of nutrients—first detected when it destroyed citrus trees more than a century ago in China —had earned a place, along with anthrax and the Ebola virus, on the Agriculture Department's list of potential agents of bioterrorism. Asian citrus psyllids, the insects that suck the bacteria out of one tree and inject them into another as they feed on the sap of their leaves, can carry the germ a mile without stopping, and the females can lay up to eight hundred eggs in their one-month life.

Kress's DNA candidate would have to fight off the bacteria or the insect. As for public acceptance, he told his industry colleagues, "We can't think about that right now."

The "Creep Factor"

Trim, silver-haired, and described by colleagues as tightly wound (he prefers "focused"), Kress arrives at the office by 6:30 each morning and microwaves a bowl of oatmeal. He stocks his office cabinet with cans of peel-top Campbell's chicken soup, which he heats up for lunch. Arriving home each evening, he cuts a rose from his garden for his wife. Weekends he works in his yard and pores over clippings about GMOs in the news.

For a man who takes pleasure in routine, the uncertainty that marked his DNA quest was disquieting. It would cost Southern Gardens millions of dollars just to perform the safety tests for a single gene in a single variety of orange. Of his five researchers' approaches, he had planned to narrow the field to the one that worked best over time.

But in 2010, with the disease spreading faster than anyone anticipated, the factor that came to weigh most was which could be ready first.

To fight *C. liberibacter,* Dean Gabriel at the University of Florida had chosen a gene from a virus that destroys bacteria as it replicates itself. Though such viruses, called bacteriophages ("phage" means to devour), are harmless to humans, Irey sometimes urged Kress to consider the public relations hurdle that might come with such a strange-sounding source of the DNA. "A gene from a virus," he would ask pointedly, "that infects bacteria?"

But Kress's chief concern was that Gabriel was taking too long to perfect his approach.

A second contender, Erik Mirkov of Texas A&M University, was further along with trees he had endowed with a gene from spinach —a food, he reminded Kress, that "we give to babies." The gene, which exists in slightly different forms in hundreds of plants and animals, produces a protein that attacks invading bacteria.

Even so, Mirkov faced skepticism from growers. "Will my juice taste like spinach?" one asked.

"Will it be green?" wondered another.

"This gene," he invariably replied, "has nothing to do with the color or taste of spinach. Your body makes very similar kinds of proteins as part of your own defense against bacteria."

When some of the scientist's promising trees got sick in their first trial, Kress agreed that he should try to improve on his results in a new generation of trees, by adjusting the gene's placement. But transgenic trees, begun as a single cell in a petri dish, can take two years before they are sturdy enough to place in the ground and many more years to bear fruit.

"Isn't there a gene," Kress asked Irey, "to hurry up Mother Nature?"

For a time, the answer seemed to lie with a third scientist, William O. Dawson at the University of Florida, who had managed to alter fully grown trees by attaching a gene to a virus that could be inserted by way of a small incision in the bark. Genes transmitted that way would eventually stop functioning, but Kress hoped to use it as a stopgap measure to ward off the disease in the 60 million citrus trees already in Florida's groves. Dawson joked that he hoped at least to save the grapefruit, whose juice he enjoyed, "preferably with a little vodka in it."

But his most promising result that year was doomed from the beginning: of the dozen bacteria-fighting genes he had then tested on his greenhouse trees, the one that appeared effective came from a pig.

One of about 30,000 genes in the animal's genetic code, it was, he ventured, "a pretty small amount of pig."

"There's no safety issue from our standpoint—but there is a certain creep factor," an Environmental Protection Agency official observed to Kress, who had included it on an early list of possibilities to run by the agency.

"At least something is working," Kress bristled. "It's a proof of concept."

A similar caution dimmed his hopes for the timely approval of a synthetic gene, designed in the laboratory of a fourth scientist, Jesse Jaynes of Tuskegee University. In a simulation, Jaynes's gene consistently vanquished the greening bacteria. But the burden of proving a synthetic gene's safety would prolong the process. "You're going to get more questions," Kress was told, "with a gene not found in nature."

In the fall of 2010, an onion gene that discouraged psyllids from landing on tomato plants was working in the Cornell laboratory of Kress's final hope, Herb Aldwinckle. But it would be some time before the gene could be transferred to orange trees.

Only Mirkov's newly fine-tuned trees with the spinach gene,

Kress and Irey agreed, could be ready in time to stave off what many believed would soon be a steep decline in the harvest. In the fall of 2010, they were put to the test inside a padlocked greenhouse stocked with infected trees and psyllids.

The Monsanto Effect

Kress's only direct brush so far with the broader battle raging over genetically modified food came in December 2010, in readers' comments on a Reuters article alluding to Southern Gardens' genetic engineering efforts.

Some readers vowed not to buy such "frankenfood." Another attributed a rise in allergies to genetic engineering. And dozens lambasted Monsanto, the St. Louis–based company that dominates the crop biotechnology business, which was not even mentioned in the article.

"If this trend goes on, one day, there will be only Monsanto engineered foods available," read one letter warning of unintended consequences.

Kress was unperturbed. Dozens of long-term animal feeding studies had concluded that existing GMOs were as safe as other crops, and the National Academy of Sciences, the World Health Organization, and others had issued statements to the same effect.

But some of his researchers worried that the popular association between GMOs and Monsanto—and in turn between Monsanto and the criticisms of modern agriculture—could turn consumers against Southern Gardens' transgenic oranges.

"The article doesn't say 'Monsanto' anywhere, but the comments are all about Monsanto," Mirkov said.

It had not helped win hearts and minds for GMOs, Kress knew, that the first such crop widely adopted by farmers was the soybean engineered by Monsanto with a bacteria gene—to tolerate a weed killer Monsanto also made.

Starting in the mid-1990s, soybean farmers in the United States overwhelmingly adopted that variety of the crop, which made it easier for them to control weeds. But the subsequent broader use of the chemical—along with a distaste for Monsanto's aggressive business tactics and a growing suspicion of a food system driven by corporate profits—combined to forge a consumer backlash. Environmental activists vandalized dozens of field trials and protested brands that used Monsanto's soybeans or corn, introduced soon after, which was engineered to prevent pests from attacking it.

In response, companies including McDonald's, Frito-Lay, and Heinz pledged not to use GMO ingredients in certain products, and some European countries prohibited their cultivation.

Some of Kress's scientists were still fuming about what they saw as the lost potential for social good hijacked by both the activists who opposed genetic engineering and the corporations that failed to convince consumers of its benefits. In many developing countries, concerns about safety and ownership of seeds led governments to delay or prohibit cultivation of needed crops: Zambia, for instance, declined shipments of GMO corn even during a 2002 famine.

"It's easy for someone who can go down to the grocery store and buy anything they need to be against GMOs," said Jaynes, who faced such barriers with a high-protein sweet potato he had engineered with a synthetic gene.

To Kress in early 2011, any comparison to Monsanto—whose large blocks of patents he had to work around and whose thousands of employees worldwide dwarfed the 750 he employed in Florida at peak harvest times—seemed far-fetched. If it was successful, Southern Gardens would hope to recoup its investment by charging a royalty for its trees. But its business strategy was aimed at saving the orange crop, whose total acreage was a tiny fraction of the crops the major biotechnology companies had pursued.

He urged his worried researchers to look at the early success of Flavr Savr tomatoes. Introduced in 1994 and engineered to stay fresh longer than traditional varieties, they proved popular enough that some stores rationed them, before business missteps by their developer ended their production.

And he was no longer alone in the pursuit of a genetically modified orange. Citrus growers were collectively financing research into a greening-resistant tree, and the Agriculture Department had also assigned a team of scientists to it. Any solution would have satisfied Kress. Almost daily, he could smell the burning of infected trees, which mingled with orange-blossom sweetness in the grove just beyond Southern Gardens' headquarters.

A Growing Urgency

In an infection-filled greenhouse where every nontransgenic tree had showed symptoms of disease, Mirkov's trees with the spinach gene had survived unscathed for more than a year. Kress would soon have three hundred of them planted in a field trial. But in the spring of 2012, he asked the Environmental Protection Agency, the first of three federal agencies that would evaluate his trees, for guidance. The next step was safety testing. And he felt that it could not be started fast enough.

Mirkov assured him that the agency's requirements for animal tests to assess the safety of the protein produced by his gene, which bore no resemblance to anything on the list of known allergens and toxins, would be minimal.

"It's spinach," he insisted. "It's been eaten for centuries."

Other concerns weighed on Kress that spring: growers in Florida did not like to talk about it, but the industry's tripling of pesticide applications to kill the bacteria-carrying psyllid was, while within legal limits, becoming expensive and worrisome. One widely used pesticide had stopped working as the psyllid evolved resistance, and Florida's citrus growers' association was petitioning one company to lift the twice-a-season restrictions on spraying young trees —increasingly its only hope for an uninfected harvest.

Others in the industry who knew of Kress's project were turning to him. He agreed to speak at the fall meeting of citrus growers in California, where the greening disease had just been detected. "We need to hear about the transgenic solution," said Ted Batkin, the association's director. But Kress worried that he had nothing to calm their fears.

And an increasingly vocal movement to require any food with genetically engineered ingredients to carry a "GMO" label had made him uneasy.

Supporters of one hotly contested California ballot initiative argued for labeling as a matter of consumer rights and transparency—but their advertisements often implied that the crops were a hazard: one pictured a child about to take a joyful bite of a pest-resistant cob of corn, on which was emblazoned a question mark and the caption "Corn, engineered to grow its own pesticide."

Yet the gene that makes corn insect-resistant, he knew, came from the same soil bacterium long used by organic food growers as a natural insecticide.

Arguing that the Food and Drug Administration should require labels on food containing GMOs, one leader of the Environmental Working Group, an advocacy group, cited "pink slime, deadly melons, tainted turkeys and BPA in our soup."

Kress attributed the labeling campaigns to the kind of tactic any industry might use to gain a competitive edge: they were financed largely by companies that sell organic products, which stood to gain if packaging implying a hazard drove customers to their own non-GMO alternatives. He did not aim to hide anything from consumers, but he would want them to understand how and why his oranges were genetically engineered. What bothered him was that a label seemed to lump all GMOs into one stigmatized category.

And when the EPA informed him in June 2012 that it would need to see test results for how large quantities of spinach protein affected honeybees and mice, he gladly wrote out the $300,000 check to have the protein made.

It was the largest single expense yet in a project that had so far cost more than $5 million. If these tests raised no red flags, he would need to test the protein as it appears in the pollen of transgenic orange blossoms. Then the agency would want to test the juice.

"Seems excessive," Mirkov said.

But Kress and Irey shared a sense of celebration. The path ahead was starting to clear.

Rather than wait for Mirkov's three hundred trees to flower, which could take several years, they agreed to try to graft his spinach-gene shoots to mature trees to hasten the production of pollen—and, finally, their first fruit—for testing.

Wall of Opposition

Early one morning a year ago, Kress checked the Agriculture Department's web site from home. The agency had opened its sixty-day public comment period on the trees modified to produce "Arctic apples" that did not brown.

His own application, he imagined, would take a similar form.

He skimmed through the company's 163-page petition, showing how the apples are equivalent in nutritional content to normal apples, how remote was the likelihood of cross-pollination with other apple varieties, and the potentially bigger market for a healthful fruit.

Then he turned to the comments. There were hundreds. And they were almost universally negative. Some were from parents, voicing concerns that the nonbrowning trait would disguise a rotten apple—though transgenic apples rotten from infection would still turn brown. Many wrote as part of a petition drive by the Center for Food Safety, a group that opposes biotechnology.

"Apples are supposed to be a natural, healthy snack," it warned. "Genetically engineered apples are neither."

Others voiced a general distrust of scientists' guarantees: "Too many things were presented to us as innocuous and years later we discovered it was untrue," wrote one woman. "After two cancers I don't feel like taking any more unnecessary risks."

Many insisted that should the fruit be approved, it ought to be labeled.

That morning Kress drove to work late. He should not be surprised by the hostility, he told himself.

Irey tried to console him with good news: the data on the honeybees and mice had come back. The highest dose of the protein the EPA wanted tested had produced no ill effect.

But the magnitude of the opposition had never hit Kress so hard. "Will they believe us?" he asked himself for the first time. "Will they believe we're doing this to eliminate chemicals and we're making sure it's safe? Or will they look at us and say, 'That's what they all say'?"

The major brands were rumored to be looking beyond Florida for their orange juice—perhaps to Brazil, where growers had taken to abandoning infected groves to plant elsewhere. Other experiments that Kress viewed as similar to his own had foundered. Pigs engineered to produce less-polluting waste had been euthanized after their developer at a Canadian university had failed to find investors. A salmon modified to grow faster was still awaiting FDA approval. A study pointing to health risks from GMOs had been discredited by scientists but was contributing to a sense among some consumers that the technology is dangerous.

And while the California labeling measure had been defeated, it had spawned a ballot initiative in Washington State and legislative proposals in Connecticut, Missouri, New Mexico, Vermont, and many other states.

In the heat of last summer, Kress gardened more savagely than his wife had ever seen.

Driving through the Central Valley of California last October to speak at the California Citrus Growers meeting, Kress considered how to answer critics. Maybe even a blanket "GMO" label would be OK, he thought, if it would help consumers understand that he had nothing to hide. He could never prove that there were no risks to genetically modifying a crop. But he could try to explain the risks of not doing so.

Southern Gardens had lost 700,000 trees trying to control the disease, more than a quarter of its total. The forecast for the coming spring harvest was dismal. The approval to use more pesticide on young trees had come through that day. At his hotel that night, he slipped a new slide into his standard talk.

On the podium the next morning, he talked about the growing use of pesticides: "We're using a lot of chemicals, pure and simple," he said. "We're using more than we've ever used before."

Then he stopped at the new slide. Unadorned, it read "Consumer Acceptance." He looked out at the audience.

What these growers wanted most, he knew, was reassurance that he could help them should the disease spread. But he had to warn them: "If we don't have consumer confidence, it doesn't matter what we come up with."

Planting

One recent sunny morning, Kress drove to a fenced field, some distance from his office and far from any other citrus tree. He unlocked the gate and signed in, as required by Agriculture Department regulations for a field trial of a genetically modified crop.

Just in the previous few months, Whole Foods had said that because of customer demand it would avoid stocking most GMO foods and require labels on them by 2018. Hundreds of thousands of protesters around the world had joined in a "March Against Monsanto"—and the Agriculture Department had issued its final report for this year's orange harvest showing a 9 percent decline from last year, attributable to citrus greening.

But visiting the field gave him some peace. In some rows were the trees with no new gene in them, sick with greening. In others were the three hundred juvenile trees with spinach genes, all healthy. In the middle were the trees that carried his immediate hopes: fifteen mature Hamlins and Valencias, 7 feet tall, onto which had been grafted shoots of Mirkov's spinach-gene trees.

There was good reason to believe that the trees would pass the EPA's tests when they bloom next spring. And he was gathering the data the Agriculture Department would need to ensure that the trees posed no risk to other plants. When he had fruit, the Food and Drug Administration would compare its safety and nutritional content to conventional oranges.

In his office is a list of groups to contact when the first GMO fruit in Florida is ready to pick: environmental organizations, consumer advocates, and others. Exactly what he would say when he finally contacted them, he did not know. Whether anyone would drink the juice from his genetically modified oranges, he did not know.

But he had decided to move ahead.

Late this summer he will plant several hundred more young trees with the spinach gene, in a new greenhouse. In two years, if he wins regulatory approval, they will be ready to go into the ground. The trees could be the first to produce juice for sale in five years or so.

Whether it is his transgenic tree or someone else's, he believed, Florida growers will soon have trees that could produce juice without fear of its being sour or in short supply.

For a moment, alone in the field, he let his mind wander.

"Maybe we can use the technology to improve orange juice," he could not help thinking. "Maybe we can find a way to have oranges grow year-round, or get two for every one we get now on a tree."

Then he reined in those thoughts.

He took the clipboard down, signed out, and locked the gate.

ROBIN MARANTZ HENIG

A Life-or-Death Situation

FROM *The New York Times Magazine*

IF MARGARET PABST BATTIN hadn't had a cold that day, she would have joined her husband, Brooke Hopkins, on his bike ride. Instead Peggy (as just about everyone calls her) went to two lectures at the University of Utah, where she teaches philosophy and writes about end-of-life bioethics. Which is why she wasn't with Brooke the moment everything changed.

Brooke was cycling down a hill in City Creek Canyon in Salt Lake City when he collided with an oncoming bicycle around a blind curve, catapulting him onto the mountain path. His helmet cracked just above the left temple, meaning that Brooke fell directly on his head, and his body followed in a grotesque somersault that broke his neck at the top of the spine. He stopped breathing, turned purple, and might have died if a flight-rescue nurse hadn't happened to jog by. The jogger resuscitated and stabilized him, and someone raced to the bottom of the canyon to call 911.

If Peggy had been there and known the extent of Brooke's injury, she might have urged the rescuers not to revive him. Brooke had updated a living will the previous year, specifying that should he suffer a grievous illness or injury leading to a terminal condition or vegetative state, he wanted no procedures done that "would serve only to unnaturally prolong the moment of my death and to unnaturally postpone or prolong the dying process." But Peggy wasn't there, and Brooke, who had recently retired as an English professor at the University of Utah, was kept breathing with a hand-pumped air bag during the ambulance ride to University Hospital, 3 miles away. As soon as he got there, he was attached to a ventilator.

By the time Peggy arrived and saw her husband ensnared in the life-sustaining machinery he had hoped to avoid, decisions about intervention already had been made. It was November 14, 2008, late afternoon. She didn't know yet that Brooke would end up a quadriplegic, paralyzed from the shoulders down.

Suffering, suicide, euthanasia, a dignified death—these were subjects she had thought and written about for years, and now, suddenly, they turned unbearably personal. Alongside her physically ravaged husband, she would watch lofty ideas be trumped by reality—and would discover just how messy, raw, and muddled the end of life can be.

In the weeks after the accident, Peggy found herself thinking about the title character in Tolstoy's *Death of Ivan Ilyich,* who wondered, "What if my whole life has been wrong?" Her whole life had involved writing "wheelbarrows full" of books and articles championing self-determination in dying. And now here was her husband, a plugged-in mannequin in the ICU, the very embodiment of a right-to-die case study.

An international leader in bioethics, Peggy explored the right to a good and easeful death by their own hand, if need be, for people who were terminally ill, as well as for those whose lives had become intolerable because of chronic illness, serious injury, or extreme old age. She didn't shy away from contentious words like "euthanasia." Nor did she run from fringe groups like NuTech, which is devoted to finding more efficient methods of what it calls self-deliverance, or SOARS (Society for Old Age Rational Suicide), which defends the right of the "very elderly" to choose death as a way to preempt old-age catastrophes. She also found common purpose with more mainstream groups, like Compassion and Choices, that push for legislation or ballot initiatives to allow doctors to help "hasten death" in the terminally ill (which is now permitted, with restrictions, in Oregon, Washington, Montana, and Vermont). And she testified in trials on behalf of individuals seeking permission to end their lives legally with the help of a doctor or a loved one.

At the heart of her argument was her belief in autonomy. "The competent patient can, and ought to be accorded the right to, determine what is to be done to him or her, even if . . . it means he or she will die," she wrote in 1994 in *The Least Worst Death,* the third of her seven books about how we die.

Peggy traces her interest in death to her mother's difficult one, from liver cancer, when Peggy was twenty-one. Only later, when, in order to write fiction, she started in an MFA program at the University of California, Irvine (which she completed while getting her doctorate in philosophy and raising two young children) did she realize how much that event had shaped her thinking. Her short stories "all looked like bioethics problems," she says, wrestling with topics like aging, mental competence, medical research, suicide—moral quandaries she would be mining for the rest of her life.

Fiction allowed her to riff on scenarios more freely than philosophy did, so she sometimes used it in her scholarly writing. In *Ending Life: Ethics and the Way We Die,* published in 2005, she included two short stories: a fictional account of an aged couple planning a tandem suicide to make way for the younger generation, until one of them has a change of heart; and a story based on an actual experience in grad school, when Peggy had to help a scientist kill the dogs in his psych experiment. The point of including the second story, she wrote in the book's introduction, was to ground her philosophical arguments in something more elemental, "the unsettling, stomach-disturbing, conscience-trying unease" of being involved in any death, whether through action, as happened in that laboratory, or acquiescence.

When Peggy finished her doctorate in 1976, the right-to-die debate was dominated by the media spectacle around Karen Ann Quinlan, a comatose young woman whose parents went to the New Jersey Supreme Court for permission to withdraw her from life support. It helped Peggy clarify her thoughts about death with dignity and shaped her belief in self-determination as a basic human right. "A person should be accorded the right to live his or her life as they see fit (provided, of course, that this does not significantly harm others), and that includes the very end of their life," she wrote in one of her nearly forty journal articles on this subject. "That's just the way I see it."

That's the way she saw it after Brooke's accident too, but with a new spiky awareness of what it means to choose death. Scholarly thought experiments were one thing, but this was a man she adored—a man with whom she had shared a rich and passionate life for more than thirty years—who was now physically devastated but still free, as she knew he had to be, to make a choice that would cause her anguish.

"It is not just about terminally ill people in general in a kind of abstract way now," she wrote after the accident; "it's also about my husband, Brooke. I still love him, that's a simple fact. What if he wanted to die? Can I imagine standing by while his ventilator was switched off?"

Before the collision, Brooke was known for his gusto. "At parties he was the one who ate the most, drank the most, talked the loudest, danced the longest," one friend recalls. A striking six-foot-five, he had a winning smile and a mess of steely gray hair and was often off on some adventure with friends. He went on expeditions to the Himalayas, Argentina, Chile, China, Venezuela, and more; closer to home, he often cycled, hiked, or backcountry skied in the mountains around Salt Lake City. In addition, Brooke, who had a bachelor's degree and a doctorate from Harvard, was a popular English professor who taught British and American literature and had a special fondness for the poetry of Wordsworth, Shelley, Byron, and Keats.

All that energy went absolutely still at the moment of his collision. When Brooke woke up in the ICU, his stepson, Mike, was at the bedside and had to tell Brooke that he might never again walk, turn over, or breathe on his own. Brooke remained silent —he was made mute by the ventilation tube down his throat—but he thought of Keats:

> The feel of not to feel it,
> When there is none to heal it
> Nor numbed sense to steel it.

"Those words, 'the feel of not to feel it,' suddenly meant something to me in ways that they never had before," he wrote later on a blog that his stepdaughter, Sara, started to keep people apprised of his progress. "My suffering was going to be a drop in the bucket compared to all the human suffering experienced by people throughout human history, but still, it was going to be a suffering nevertheless."

Brooke took some solace in Buddhism, which he began exploring when he was in his forties. A few weeks after the accident, a local Buddhist teacher, Lama Thupten Dorje Gyaltsen, came to his hospital room. "The body is ephemeral," Lama Thupten declared, gesturing at his own body under his maroon-and-saffron robe. He urged Brooke to focus on his mind. At the time, it was a comfort to think that his mind, which seemed intact, was all that mattered. It meant he could still be the same man he always was even if he never moved again. But as much as he yearned to believe it, Brooke's subsequent experiences—spasms, pain, catheterizations, bouts of pneumonia, infected abscesses in his groin—have made him wary of platitudes. He still wants to believe the mind is everything. But he has learned that no mind can fly free of a useless body's incessant neediness.

One gray morning in February, more than four years after the accident, I met Brooke and Peggy at their home in the Salt Lake City neighborhood known as the Avenues. Brooke rolled into the living room in his motorized wheelchair. It was a month before his seventy-first birthday, and his handsome face was animated by intense, shiny brown eyes, deep-set under a bristly awning of brow. He was dressed as usual: a pullover, polyester pants that snap open all the way down each leg, a diaper, and green Crocs. A friend was reading on a couch nearby, a caregiver was doing her schoolwork in the kitchen, and Peggy had retreated upstairs to her office amid towers of papers, books, and magazines. She had finally gained some momentum on a project that was slowed by Brooke's accident: a compendium of philosophical writings about suicide, dating as far back as Aristotle.

Peggy, who is seventy-two, still works full-time. This lets her hold on to the university's excellent health insurance, which covers a large portion of Brooke's inpatient care and doctor bills, with Medicare paying most of the rest of them. But even with this double coverage, Peggy spends a lot of time arguing with insurance companies that balk at expenditures like his $45,000 wheelchair. And she still pays a huge amount of the cost, including nearly $250,000 a year to Brooke's caregivers, twelve mostly young and devoted health care workers, who come in shifts so there's always at least one on duty. Peggy says she and Brooke were lucky to have had a healthy retirement fund at the time of the accident, but she doesn't know how many more years they will be able to sustain this level of high-quality twenty-four-hour care.

Scattered around the living room were counter-height stools that Peggy picked up at yard sales. She urges visitors to pull them up to Brooke's wheelchair, because he's tall and the stools bring most people to eye level. About two years ago, Brooke used a ventilator only when he slept, but following a series of infections and other setbacks, he was now on the ventilator many of his waking hours too, along with a diaphragmatic pacer that kept his breathing regular. Earlier that morning his caregiver had adjusted the ventilator so he and I could talk, deflating the cuff around his tracheostomy tube to allow air to pass over his larynx. This let him speak the way everyone does, vocalizing as he exhaled. It seemed to tire him, though; his pauses became longer as our conversation went on. But whenever I suggested that we stop for a while so he could rest, Brooke insisted that he wanted to keep talking.

What he wanted to talk about was how depressed he was. He recognized the feeling, having struggled with bipolar disorder since adolescence. "It takes a long time to get ready for anything," he said about his life now. "To get up in the morning, which I kind of hate, to have every day be more or less the same as every other day . . . and then to spend so much time going to bed. Day after day, day after day, day after day."

Brooke has good days and bad days. When friends are around playing blues harmonica or reading aloud to him, when his mind is clear and his body is not in pain—that's a good day. On a good day, he said, he feels even more creative than he was in his able-bodied life, and his relationships with Peggy, his two stepchildren, and his many friends are richer and more intimate than before; he has no time or patience for small talk, and neither do they. Every so often he'll turn to Peggy and announce, "I love my life."

On a good day, Brooke's voice is strong, which lets him keep up with reading and writing with voice-recognition software. A caregiver arranges a Bluetooth microphone on his head, and he dictates e-mail and races through books by calling out "Page down" when he reaches the bottom of a screen. On a good day, he also might get outside for a while. "I like to take long walks, quote unquote, in the park," he told me. "There's a graveyard somewhat lugubriously next to us that I like to go through," pushed in his wheelchair by a caregiver with Peggy alongside. A couple of years ago, he and Peggy bought two plots there; they get a kick out of visiting their burial sites and taking in the view.

But on bad days these pleasures fade, and everything about his current life seems bleak. These are days when physical problems — latent infections, low oxygen levels, drug interactions, or, in a cruel paradox of paralysis, severe pain in his motionless limbs — can lead to exhaustion, depression, confusion, and even hallucinations. As Brooke described these darker times, Peggy came down from her office and sat nearby, half-listening. She has bright blue eyes and a pretty, freckled face fringed by blond-white hair. Most days she wears jeans and running shoes and a slightly distracted expression. She takes long hikes almost daily, and once a week tries to squeeze in a Pilates session to help treat her scoliosis. Each body harbors its own form of decay, and this is Peggy's; the scoliosis is getting worse as she ages.

She walked over to us, bent crookedly at the waist, and gently kissed Brooke's forehead. "Depression is not uncommon in winter," she said in the soft voice she almost always uses with him. "It's important to think positive thoughts."

"Basically I dislike being dependent, that's all," he said, looking hard into her eyes. He spit some excess saliva into a cup.

"It's something you never complain about," she said. "You're not a big complainer."

"One thing I don't like is people speaking for me, though."

Peggy looked a bit stung. "And that includes me?" she asked.

"Yes," he said, still looking into her eyes. "I don't like that."

She made an effort not to get defensive. "Well, sometimes that has to happen, for me to speak for you," she began. "But . . . but not always. I try not to."

Brooke seemed sorry to have spoken up; it was clear he didn't want to hurt her. "I'm trying to be as frank as possible," he said.

"No, it's good," she assured him, her protective instincts clicking in. "It helps me for you to say that, to tell me what you would have wanted to say instead."

All Brooke could muster was a raspy "Yep."

"The most important thing is to not speak for someone else," Peggy insisted.

"Yep," Brooke repeated. "What I want to do most right now is be quiet and read." So Peggy and I left him in the living room, where the big-screen monitor was queued up to Chapter 46 of *Moby-Dick*. "Page down," he called out, forced to keep repeating it like a mantra because his speech was croaky and the software had trouble recognizing the phrase. "Page down. Page down."

For Brooke, what elevates his life beyond the day-to-day slog of maintaining it—the vast team effort required to keep his inert sack of a body fed and dressed and clean and functioning—is his continuing ability to teach part-time through the University of Utah's adult education program. During my February visit, I sat in on one of his classes, which he teaches with Michael Rudick, another retired English professor from the university. Some two dozen students, most over sixty, crammed into Brooke's living room for a discussion of *Moby-Dick*. Conversation turned to the mind-body problem. "Melville is making fun here of Descartes, as though you could exist as a mind without a body," said Howard Horwitz, who teaches in the English department and was helping out that day.

Brooke seemed exhausted and sat quietly, impassive as Buddha, as his ventilator sighed. At one point a student called out to ask what Brooke thought about a particular passage. He responded with an oblique "I'd much rather hear what you think," and was silent for the rest of the class. The discussion continued with the two other professors taking charge. There was an almost forced animation, as if the students had tacitly agreed to cover for a man they loved, admired, and were worried about.

When Peggy arrived late—she was at a meeting on campus— Brooke flashed her one of his dazzling smiles. His eyes stayed on her as she positioned herself near an old baby grand that hugs a corner of the living room, a memento from Brooke's parents' house in Baltimore. Above the piano is a huge painting that Peggy got years ago, a serial self-portrait of a dark-haired figure with a mustache—six full-body images of the same man in various stages of disappearing.

"He's never looked this bad," Peggy whispered to me during the break as students milled around. She went to Brooke and kissed his forehead. "Are you OK?" she asked softly.

"I'm fine," he said. "Don't worry."

They have this exchange a lot: Peggy leaning in to ask if he's OK, Brooke telling her not to worry, Peggy worrying anyway. Quietly, so the students wouldn't hear, she asked the respiratory therapist on duty, Jaycee Carter, when Brooke last had his Cough Assist therapy, a method that forces out mucus that can clog his lungs. "Three hours ago," Jaycee said. But Brooke said he didn't want it while the class was there: it's noisy, and it brings up a lot of unsightly phlegm. As students started to head back to their seats, Peggy lit on a more discreet alternative: a spritz of albuterol, used in asthma inhalers to relax the airways, into his trach tube. Jaycee stood by awaiting instructions, Brooke kept shaking his head—no albuterol, not now, no—and Peggy kept insisting. At last, annoyance prickling his expressive eyebrows, he gave in, and Jaycee did as she was told. But the albuterol didn't help.

Peggy retreated to the piano as the class resumed, her eyes brimming. "This is bad," she murmured. "This is really bad." Underlying her anxiety was a frightening possibility: that Brooke's inability to teach that day was the start of a progressive decline. Up until then, his occasional mental fogginess was always explained by something transient, like an infection. But if he were to lose his intellectual functioning, he would be robbed of all the things that still give his life meaning: teaching, writing, and interacting with the people he loves. If that day ever came, it would provoke a grim reckoning, forcing Brooke to rethink—provided he was still capable of thinking—whether this is a life worth holding on to.

After class, Jaycee wheeled Brooke to the dining area so he could sit with Peggy and me as we ate dinner. Brooke doesn't eat anymore. Last August he had a feeding tube inserted as a way to avoid the dangerous infections and inflammations that were constantly sending him to the hospital. If he doesn't chew, drink, or swallow, there's less chance that food or fluid will end up in his lungs and cause aspiration pneumonia.

In his prior life, Brooke couldn't have imagined tolerating a feeding tube; he loved eating too much. In fact, when he updated his living will in 2007, he specifically noted his wish to avoid "administration of sustenance and hydration." But the document had a caveat found in most advance directives, one that has proved critical in negotiating his care since the accident: "I reserve the right to give current medical directions to physicians and other providers of medical services so long as I am able," even if they conflict with the living will.

Thus a man who had always taken great joy in preparing, sharing, and savoring food decided to give up his final sensory pleasure in order to go on living. He swears he doesn't miss it. He had already been limited to soft, easy-to-swallow foods with no seeds or crunchiness—runny eggs, yogurt, mashed avocado. And as much as he loved the social aspects of eating, the long conversations over the last of the wine, he managed, with some gentle prodding from Peggy, to think of the feeding tube as a kind of liberation. After all, as she explained on the family blog, Brooke could still do "almost all the important things that are part of the enjoyment of food" —he could still smell its aroma, admire its presentation, join in on the mealtime chatter, even sample a morsel the way a wine taster might, chewing it and then discreetly spitting it out. Maybe, she wrote, "being liberated from the crass bodily necessity of eating brings you a step closer to some sort of nirvana."

Or as Brooke put it to me in his unvarnished way: "You can get used to anything."

Brooke kept nodding off as he sat watching us eat—the class had really drained him—but Peggy kept him up until nine o'clock, when his hourlong bedtime ritual begins. After Jaycee took him to his room, she and the night-shift caregiver hoisted him from his wheelchair and into the bed using an elaborate system of ceiling tracks, slings, and motorized lifts; changed him into a hospital gown; washed his face and brushed his teeth; emptied his bladder with a catheter; strapped on booties and finger splints to position his extremities; hooked him up to the ventilator; and set up four cans of Replete Fiber to slowly drip into his feeding tube as he slept. The ritual ended with what Brooke and Peggy think of as the most important part of the day, when Brooke finally is settled into bed and Peggy takes off her shoes and climbs in, too, keeping him company until he gets sleepy. (Peggy sleeps in a new bedroom she had built upstairs.) There they lie, side by side in his double-wide hospital bed, their heads close on the pillow, talking in the low, private rumbles of any intimate marriage.

Throughout the first half of last year, Brooke had severe pain in his back and legs, and all the remedies he tried—acupuncture, cortisone shots, pressure-point therapy, nerve-impulse scrambling —were useless. At one point last summer, he decided he couldn't go on living that way. "Pain eats away at your soul," he wrote on July 28, 2012, using his voice-recognition software to dictate what he called a "Final Letter" to his loved ones, explaining why he now wanted to die:

> For many years since the accident I have been motivated by a deep will to live and to contribute to the benefit of others in my small way. I think I have done that. And I am proud of it. But as I have told Peggy over the past few months, I knew that I would reach a limit to what I could do. And I have arrived at the limit over the past couple of weeks.

He had had thoughts like this before, but this time it felt different to Peggy, who proofread and typed the letter; the longing for death felt like something carefully considered, something serious and sincere. This was an autonomous, fully alert person making a decision about his own final days—the very situation she had spent her career defending. She reasoned that Brooke had the right, as a mentally competent patient, to reject medical interventions that could further prolong his life, even though he did not live in a state where assisted suicide was explicitly legal. And if he wanted to reject those interventions now, after four years of consenting to every treatment, Peggy was ready to help. She shifted from being Brooke's devoted lifeline to being the midwife to his death.

She knew from a hospice nurse that one way to ease a patient's dying included morphine for "air hunger," Haldol for "delusions and end-of-life agitation," and Tylenol suppositories for "end-of-life fever, 99 to 101 degrees." Another nurse mentioned morphine, Haldol, and the sedative Ativan; a third talked about Duragesic patches to deliver fentanyl, a potent opium alternative used for pain. Peggy also tried to find out whether cardiologists would ever be willing to order deactivation of a pacemaker at a very ill patient's request (probably, she was told). She kept pages of scribbled notes in a blue folder marked "Death and Dying." She had also taken careful notes when Brooke started to talk about his funeral. He told her what music he wanted, including a few gospel songs by Marion Williams, and which readings from Wordsworth's "Lucy Poems" and Whitman's *Leaves of Grass*. On his gravestone, he might like a line from Henry Adams: "A teacher affects eternity; he can never tell where his influence stops." These were good conversations, but they left him, he told Peggy, "completely emotionally torn up."

Then in early August, fluid started accumulating in Brooke's chest cavity, a condition known as pleural effusion, and he had trouble breathing, even on the ventilator. He was uncomfortable and becoming delirious. Other people, including a few of Brooke's caregivers, might have seen this as a kind of divine intervention—a rapid deterioration just when Brooke was longing for death anyway, easing him into a final release. But that's not how Peggy saw it. This was not the death Brooke wanted, confused and in pain, she explained to me later; he had always spoken of a "generous death" for which he was alert, calm, present, and surrounded by people he loved. So she consulted with a physician at the hospital about whether Brooke would improve if doctors there extracted the fluid that was causing the respiratory distress. In the end, she decided to ignore the "Final Letter." She went upstairs, got dressed, and, along with the caregiver on duty, put Brooke into the wheelchair-accessible van in the driveway and drove him to the emergency room.

This put Brooke back in the hospital with heavy-duty antibiotics treating yet another lung problem. During his three-week stay he recovered enough to make his own medical decisions again —which is when he consented to the insertion of the feeding tube. He also met with a palliative-care expert, who suggested trying one more pain treatment: low-dose methadone around the clock, 5 milligrams at exactly 9 A.M. and exactly 9 P.M. every day. With the methadone, Brooke's pain was at last manageable. Now when he reflects on that hospitalization, he thinks of it as having a "happy ending." In the "Death and Dying" folder is one last penciled note from Peggy dated August 18, 2012: "10:37 A.M. Brooke says he wants to 'soldier on' despite difficulties."

A couple of days after Brooke and Peggy talked about his not wanting anyone to speak for him, the subject came up again. Peggy raised it as we all sat in the living room. At first she did all the talking, unwittingly acting out the very problem under discussion. So I interrupted with a direct question to Brooke. Why, I asked, do you think Peggy sometimes does the talking for you?

"I think it's because she's concerned about me and wants the best for me," he said. He made the gesture I'd watched him make before, lifting the tops of his shoulders, over which he still has motor control, in a resigned-looking little shrug. In light of such pervasive dependency, that shrug seemed to say, how can a loving, well-meaning wife help but sometimes overstep in her eagerness to anticipate her husband's needs?

I asked Brooke if Peggy ever misunderstood what he meant to say.

"I don't know, ask her," he said. But Peggy saw the irony there and urged Brooke to speak up for himself.

"Occasionally, yes," he said, though he couldn't think of any specific instances.

When she makes a mistake, I asked, do you ever correct her?

"No, because I don't want to upset her." His brown eyes got very big.

She: "It would be OK."

He: "OK."

She: "It would help me if you would say to me —"

He: "OK, OK, OK."

She: "I think this issue is especially important . . . . What you've wanted has fluctuated a lot, and part of it is to try to figure out what's genuine and what's a part of response to the pain. That's the hardest part for me, when you say: 'I don't want to go to the hospital ever again, I don't like being in the hospital, and I don't want to be sick. If the choice is going to the hospital or dying, I'll take the dying.'"

Peggy turned to me. She wanted me to understand her thinking on this. It's so hard to know what Brooke wants, she explained, because there have been times when she has taken him to the hospital, and he later says that she made the right call. It's so hard, she repeated. She has to be able to hear how a transient despair differs from a deep and abiding decision to die. She believes he hasn't made that deep, abiding decision yet, despite the "Final Letter."

She understands him well enough, she told me, to know when his apparent urgency is just a reflection of his dramatic way of presenting things: his deep voice, his massive size, his grimaces. "Brooke is very expressive when he's in his full self," she said.

Watching the dependence, indignity, and sheer physical travail that Brooke must live through every day, Peggy told me, she doesn't think she would have the stamina to endure a devastating injury like his. "It seems not what I'd want," she said when I asked if she would choose to stay alive if she were paralyzed. While she might not want to persevere in such a constrained and difficult life, she believes that Brooke does want to, and she tends to interpret even his most anguished cries in a way that lets her conclude that he doesn't quite mean what he says. But she worries that others in his life, even the caregivers who have become so close to him, might not be able to calibrate the sincerity of those over-the-top pleas and might leap too quickly to follow his instructions if he yelled out about wanting to end it all.

Suzy Quirantes, the senior member of the caregiving team, a trained respiratory therapist who has been with Brooke since the day he came home in 2010, sees it a bit differently. "I've worked with death a lot," she told me. She thinks there have been times when Peggy has been unable to hear Brooke's heartfelt expressions of a desire to die. "Last year, right after the feeding tube, he kept refusing his therapies," she said. "And I said, 'If you're really serious, if you're done, I need you to be very clear, and you need to be able to talk to Peggy so she understands.'" He never did talk to Peggy, though—maybe because he wasn't clear in his own mind what he wanted. "He has said, 'I'm done,' and then when we kind of talk more about it, he gets scared," Suzy said. "He says: 'What I mean is I'm done doing this stuff in the hospital. But I'm not ready to die yet.'"

The tangled, sometimes contradictory nature of Brooke's feelings has led to subtle shifts in Peggy's scholarly thinking. She still believes that, whenever possible, people have the right to choose when and how to die. But she now better understands how vast and terrifying that choice really is. "What has changed," she told me, "is my sense of how extremely complex, how extremely textured, any particular case is." This realization is infinitely more fraught when you're inextricably invested in the outcome and when the signals your loved one sends are not only hard to read but also are constantly in flux.

The only consistent choice Brooke has made—and he's made it again and again every time he gives informed consent for a feeding tube or a diaphragmatic pacer, every time he permits treatment of an infection or a bedsore—is the one to stay alive. This is the often unspoken flip side of the death-with-dignity movement that Peggy has long been a part of. Proponents generally focus on only one branch of the decision tree: the moment of choosing death. There's much talk of living wills, DNR orders, suicide, withdrawal of life support, exit strategies. Brooke's experience has forced Peggy to step back from that moment to an earlier one:

the moment of confronting one's own horrific circumstances and choosing, at least for now, to keep on living. But the reasons for that choice are complicated too. Brooke told me that he knows Peggy is a strong person who will recover from his death and move on. But he has also expressed a desire not to abandon her. And Peggy worries that sometimes Brooke is saying he wants to keep fighting and stay alive not because that's what he wants, but because he thinks that's what she wants him to want. And to further complicate things, it's not even clear what Peggy really wants him to want. Her own desires seem to shift from day to day. One thing doesn't change, though: she is deeply afraid of misunderstanding Brooke's wishes in a way that can't be undone. The worst outcome, to her, would be to think that this time he really did want to die and then to feel as if she might have been wrong.

Since Brooke's accident, Peggy has continued to advocate for people seeking to die. She went to Vancouver in late 2011 to testify in court in the case of Gloria Taylor, a woman with ALS who wanted help ending her life when she was ready. And in 2012 she presented testimony by Skype in the case of Marie Fleming, an Irishwoman with multiple sclerosis who was making a similar request. The plaintiffs were a lot like Brooke, cognitively intact with progressively more useless bodies. But they felt a need to go to court to assure that they would have control in the timing of their own deaths. Brooke has not. Perhaps that's because he believes that Peggy will follow through on a plan to help him die if that's what he ultimately chooses.

Those seeking to end their lives are up against opponents who say that helping the terminally ill to die will lead eventually to pressure being put on vulnerable people—the elderly, the poor, the chronically disabled, the mentally ill—to agree to die to ease the burden on the rest of us. Peggy doesn't buy it. The scholarly work she is most proud of is a study she conducted in 2007, which is one of the first to look empirically at whether people are being coerced into choosing to end their lives. Peggy was reassured when she and her colleagues found that in Oregon and the Netherlands, two places that allow assisted dying, the people who used it tended to be better off and more educated than the people in groups considered vulnerable.

What Peggy has become more aware of now is the possibility of the opposite, more subtle kind of coercion—not the influence of a greedy relative or a cost-conscious state that wants you to die, but pressure from a much-loved spouse or partner who wants you to live. The very presence of these loved ones undercuts the notion of true autonomy. We are social beings, and only the unluckiest of us live in a vacuum; for most, there are always at least a few people who count on us, adore us, and have a stake in what we decide. Everyone's autonomy abuts someone else's.

During Peggy's cross-examination in the Gloria Taylor trial, the Canadian government's lawyer tried to argue that Brooke's choice to keep living weakened Peggy's argument in favor of assisted suicide. Isn't it true, the lawyer asked, that "this accident presented some pretty profoundly serious challenges to your thinking on the subject?"

Yes, Peggy said, but only by provoking the "concerted re-re-rethinking" that any self-respecting philosopher engages in. She remained committed to two moral constructs in end-of-life decision making: autonomy and mercy. "Only where both are operating —that is, where the patient wants to die and dying is the only acceptable way for the patient to avoid pain and suffering—is there a basis for physician-assisted dying," she told the court in an affidavit. "Neither principle is sufficient in and of itself, and in tandem the two principles operate as safeguards against abuse."

One morning in April, I called to speak with Peggy and Brooke. Peggy told me that when I was there in February, Brooke had an undiagnosed urinary tract infection that affected both his body and his clarity of thinking. It had since cleared up, she said. "He's a different person than the one you saw." The possibility that he'd begun a true cognitive decline was averted, at least for the time being.

"I'm cautiously happy about life in general," Brooke said on speakerphone, stopping between phrases to catch his breath. "I'm getting stronger. Working hard. Loving my teaching. My friends and caregivers. My wife."

I asked about Brooke's "Final Letter" from the previous summer. I was still trying to understand why Peggy had ignored it, just days after she typed it up for him, and instead took him to the ER to treat his pleural effusion. Why hadn't she just let the infection end his life?

"Brooke had always said, 'I'm willing to go to the hospital for something that's reversible, but I don't want to die in the hospital,'" she said, as Brooke listened in on the speakerphone. So she had to "intuit" whether this was something reversible, and she believed it was. "This didn't feel like the end," she said, "but of course you don't know that for sure." In addition, there was that image in her mind of Brooke's ideal of a "generous death." It's hard to say whether she'll ever think conditions are exactly right for the kind of death Brooke wants.

The next day I learned that a few hours after my phone call, Brooke suddenly became agitated and started to yell. "Something bad is happening," he boomed. "I'm not going to make it through the morning." Peggy and the caregiver on duty, Jaycee, tried to figure out what might have brought this on, just hours after he told me he was "cautiously happy." He had gone the previous two nights without his usual Klonopin, which treats his anxiety; maybe that was the explanation. Or maybe discussing his "Final Letter" with me, remembering the desperation of that time, had upset him. He was also getting ready for the first class of a new semester, covering the second half of *Moby-Dick;* maybe he was experiencing the same teaching anxiety that had plagued him his whole career.

Deciding that Brooke was having a panic attack, Peggy told Jaycee to give him half a dose of Klonopin. She did, but things got worse. Brooke's eyes flashed with fear, and he yelled to Peggy that he was about to do something terrible to her—meaning, she guessed, that he was going to die and leave her alone. Finally he announced that he wanted to turn off all the machines. Everything. He wanted to be disconnected from all the tubes and hoses that were keeping him alive. He was ready to die.

Peggy and Jaycee did what he asked. They turned off the ventilator and disconnected it from the trach and placed a cap at the opening in his throat. They turned off the oxygen. They turned off the external battery for the diaphragmatic pacer. They showed Brooke that everything was disconnected.

Brooke sat back in his wheelchair then and closed his eyes. There were no tears, no formal goodbyes; it all happened too quickly for that. He sat there waiting to die, ready to die, and felt an incredible sense of calm.

Two minutes passed. Three minutes passed. He opened his eyes and saw Peggy and Jaycee sitting on stools, one on either side, watching him.

"Is this a dream?" he asked.

"No, it's not a dream."

"I didn't die?"

To Brooke it was a kind of miracle—all the machinery had been shut off, just as he asked, but he was still alive. He felt refreshed, as if he had made it through some sort of trial. He asked Jaycee to reattach everything, and three hours later, after he had a nap, his students arrived to start the new semester, and Brooke began teaching *Moby-Dick* again.

But it was no miracle. "I know what his medical condition is," Peggy told me later, out of Brooke's earshot. "The reason he didn't die is he's not at the moment fully vent-dependent anymore. He can go without oxygen for a while, and he can go with the pacer turned off for some time." She didn't say any of this to Brooke. "It seems to have been such an epiphany, such a discovery, when he woke up and discovered he was still alive," she said. "I don't really want to puncture that bubble."

If for some reason Brooke had become unconscious, she and Jaycee would have revived him, Peggy told me, because she didn't believe he really wanted to die. She thinks what he really wanted was to believe he had a measure of control, that he could ask for an end to his life and be heard. "We showed him that we would do what he asked for," she said, "and he thought it was real." But it wasn't real, I said. It all sounded like an elaborate end-of-life placebo, an indication that in fact he was not in control, that he wasn't being heard. Peggy laughed and did not disagree.

She's not good at keeping secrets from Brooke, though, and by the time I contacted them both by Skype later in the week, she'd told him the truth about that afternoon. In retrospect, Brooke said, the whole thing seemed kind of comical. He mimed it for me, leaning back with his eyes closed waiting for the end to come, then slowly opening them, raising his eyebrows practically to his hairline, overacting like a silent-film star tied to the tracks who slowly realizes the distant train will never arrive. He looked good, handsome in his burgundy polo shirt, mugging for the webcam. Some new crisis, some new decision, was inevitable—in fact, last month it took the form of another farewell letter, stating his desire to die in the spring of 2014, which is when he expects to be finished teaching his next course, on *Don Quixote*. But at that moment, Brooke was feeling good. "I think it will be a productive summer," he said. And he and Peggy smiled.

VIRGINIA HUGHES

23 and You

FROM *Matter*

CHERYL WHITTLE TRIED HER BEST to fall asleep, but her mind kept racing. Tomorrow was going to be the culmination of three years of research and, possibly, a day that would change her life forever.

Around 4 A.M. she popped two Benadryl and managed to drift off. But in just a few hours she had to be up and ready to go.

Cheryl and her husband, Dickie, are retired and live in eastern Virginia, way out on the end of the Northern Neck peninsula, which juts like an arthritic finger into Chesapeake Bay. It's a beautiful and isolated spot, where most people tack up NO TRESPASSING signs and stay close to home. The Whittles enjoy their life in the country, but Cheryl was eager that day to make the long drive to meet Effie Jane.

She showered, threw on a T-shirt, jeans, and sneakers, dotted makeup on her cheeks, and scrunched a dollop of mousse into her thinning brown hair.

There's nothing showy about Cheryl, not even on a day like this. She's short and shy, with nine grandchildren and no pretensions. She grabbed a shoulder bag, heavy with the day's supplies, and kissed Dickie on her way out the door.

Her anxiety mounted as she drove her yellow pickup truck past sleepy cornfields, old plantations, and cemeteries, up the peninsula and into mainland Virginia. Then she pulled into the tiny parking lot of Panera Bread in Richmond. She didn't have to wait

Note: Some names have been changed to protect the privacy of individuals.

long before Effie Jane Erhardt found her—that yellow truck was hard to miss. Effie Jane pulled open the truck's passenger door and announced, "I'm here!"

Cheryl and Effie Jane met on Ancestry.com, a popular web site for people trying to fill in their family trees. After several e-mail and phone encounters, each woman felt a kinship that neither had experienced before. Both were born in 1951, and grew up about 20 miles from each other in the Richmond area. They both speak with soft southern drawls, had traumatic childhoods, are devout Christians, and, as children, felt like outsiders in their own families.

Cheryl quickly got down to business, retrieving a small cardboard box from her bag in the back seat. She opened the top, plucked out a fat plastic tube, and handed it to her friend. Effie Jane held the tube under her mouth and spit—and spit, and spit, and spit. She had never realized how much saliva froths and fizzes. She passed the tube back to Cheryl, who snapped on a plastic cap, gently mixed the tube's contents, and dropped it into a clear plastic bag with an orange BIOHAZARD label. Then the two women went into Panera for lunch.

Since 2000, when a company called Family Tree DNA sold the first commercially available home testing kit, an estimated one million people have dabbled in genetic genealogy—also known as recreational genetics, extreme genealogy, and even anthrogenealogy.

Traditionally, amateur genealogical research was regarded as a niche hobby for older white men, but today it attracts people of all ages, races, and walks of life.

The rapid transformation is the result of two technological revolutions. Twenty years ago, doing genealogy meant hitting the pavement: traveling to local historical societies, courthouses, libraries, and cemeteries to paw through dusty books and records. Then came the Internet, which made the most useful references —census and voter lists, birth certificates, military records, even the archives of local newspapers—accessible from home. Not only that, but genealogists started connecting with each other online, sharing their research and overlapping trees, creating a vast online database that anyone could tap into and, more importantly, add to.

In 1997 a company called Infobases, which sold compact disks of the Church of Jesus Christ of Latter-day Saints publications, bought *Ancestry Magazine* and its web site, Ancestry.com, turning the latter into a subscription genealogy service. By 2009, when Ancestry.com went public, it had a near monopoly on the booming industry. The world of ancestry research has become a perfect example of a highly scalable business based largely on freely provided, user-generated content. Today Ancestry.com has a few competitors, like MyHeritage.com and Brightsolid, but it remains dominant, with almost 3 million paying subscribers, 12 billion records, and 50 million family trees. Revenues from the company's ten popular web sites and the Family Tree Maker software totaled $400 million in 2011. In late 2012 a European private equity firm bought the company for $1.6 billion.

The second transformation came from rapid advances in genetic testing. After Family Tree DNA launched its test, other companies followed: eleven by 2004, and almost forty by the end of that decade. Today you find celebrities like Meryl Streep and Yo-Yo Ma tracing their lineage on prime-time television shows. As the price of commercial genetic tests has plummeted—many now cost just $99—families like the Whittles have been able to join in.

Three companies—23andMe, Family Tree DNA, and Ancestry .com—have emerged as major players, and each is intent on growing its most valuable asset: a proprietary database of customers' genetic data. 23andMe has information from more than 400,000 people and counting, and Family Tree DNA has over 650,000 different genetic records.

The bigger these databases become, the more useful they are for filling in genealogists' ever-expanding family trees. But the growth of the databases also raises serious privacy concerns—not only for people who buy the tests, but for close or even distant family members who share some of their DNA.

Searching your genetic ancestry can certainly be fun: you can trace the migration patterns of 10,000-year-old ancestors or discover whether a distant relative ruled a continent or rode on the *Mayflower.* But the technology can just as easily unearth more private acts—infidelities, sperm donations, adoptions—of more recent generations, including previously unknown behaviors of your grandparents, parents, and even spouses. Family secrets have never been so vulnerable.

If you find a relative on a genetic genealogy database—say, a second or third cousin—then, with the help of Google, social media, digital obituaries, and other publicly available resources, it's usually possible to find closer kin. Adoptees have used their newfound genetic knowledge to browse photo albums and look for potential biological relatives on Facebook. Children of sperm donors have found siblings they never knew they had. Couples who used artificial insemination to conceive have discovered that another man's sperm was used.

And then there are people like Cheryl, who learn to their surprise, late in life, that they aren't the person they thought they were.

Over sandwiches at Panera, Cheryl and Effie Jane exchanged photos and told childhood stories. Taking advantage of the restaurant's Wi-Fi, Cheryl took out her laptop and logged in to the 23andMe web site, patiently explaining how the process worked. Cheryl was a veteran. Genetic testing had already shaken up her world, raising startling new questions about where she came from. She was here because she believed that Effie Jane was her sister. She was praying for it. If she was right, the journey she'd been on for the last three years would reach its end. Her mind could rest.

The women left the restaurant together, drove to a nearby post office, and sent the sealed package to a lab in Los Angeles. There technicians would screen Effie Jane's DNA for about one million genetic markers. Four to six weeks later, 23andMe would send Cheryl an e-mail saying the results were ready.

Cheryl's quest began one afternoon in late 2008, when she and Dickie were sitting in their living room watching *Oprah*. The episode included a segment about a Silicon Valley start-up called 23andMe that was selling genetic tests directly to consumers. One of the company's founders, Anne Wojcicki, was nine months pregnant. She told Oprah how her DNA test results and those of her husband—Google cofounder Sergey Brin—offered clues about their unborn child.

Cheryl, then a registered nurse, was intrigued. She had married Dickie when she was fourteen and he was twenty. The pregnancy that spurred their young union resulted in a stillborn girl, born with too much fluid in her brain.

After hearing Wojcicki's story, Cheryl thought that this DNA test might provide a genetic explanation for their daughter's death. What's more, she had been interested in genealogy for years and had done a lot of work on the line of her father, Josiah "Joe" Wilmoth. Perhaps DNA testing would expand that research.

Cheryl remembered some of the basics about how genes work from nursing school, and learned more after browsing 23andMe's web site. For instance, most of us have twenty-three pairs of chromosomes, or long segments of DNA. Both sexes have twenty-two pairs of so-called "autosomal chromosomes"—each pair includes one copy inherited from each parent. But the twenty-third pair, the "sex chromosomes," is different. Men inherit a Y chromosome from their father and an X chromosome from their mother. Women, in contrast, receive an X chromosome from their father and a second X from their mother.

To investigate her father's ancestry, Cheryl decided to look at the DNA of a male in her father's line. Joe Wilmoth had died in 1989. But his son Milton Wilmoth, Cheryl's older half-brother, was alive and well. So she called Milton and asked: If she paid for it, would he consider taking a genetic test?

Milton didn't own a computer and had no interest in his DNA, but he readily agreed to help his sister. So in early 2009 Cheryl bought three kits from 23andMe—one for her, one for her husband Dickie, and one for Milton—at $495 a pop. Two months after that, she was back at her computer poring over the results.

From what she could tell, nothing in her genes or Dickie's gave any clues about why their first baby had died. But not long after she took the test, 23andMe launched Relative Finder, a service that allowed customers to find relatives in the company's database based on shared segments from all twenty-three chromosomes. Oddly, Milton's name did not appear in Cheryl's list of DNA relatives. She tried signing in to 23andMe using Milton's account instead, and saw that her name did not appear in his list of DNA relatives either.

After a couple of months, Cheryl reached out to CeCe Moore, an expert in genetic genealogy who runs a popular blog on the subject. "I wrote to her and said, 'Can you tell me what I'm doing wrong? I can't get this thing to work,'" Cheryl recalls.

CeCe revealed the truth that Cheryl suspected but had been scared to confront: Milton was not biologically related to her. For

Cheryl, there was only one explanation: she was a Wilmoth by name but not by blood.

The notion that Joe wasn't her biological father didn't sit well with Cheryl. It meant, after all, that her mother had lied to her —and maybe to Joe and everybody else—for decades. But Cheryl didn't feel anger toward her mother. In fact, the more she thought about it, the more she felt comforted. Many things about her childhood suddenly made sense. When she was growing up, Joe never gave her much affection or attention. And when he drank, which was often, he could turn mean. "I never felt a part of him," she says. "I grew up believing in my heart that I did not belong."

The news was a painful shock, however, to Cheryl's younger sister, Sandi Satterfield, who was crushed at the thought that they weren't full siblings. Joe had always doted on Sandi, and Sandi had adored him, despite his flaws. Yet she worried about what it might mean for her own roots. Could it possibly be that she, too, wasn't his?

After prodding from Cheryl, Sandi agreed to take a 23andMe test to confirm that Joe was her father. When her results came back, in June 2010, they showed that she was at extremely high risk for colorectal cancer. Just seven months later she was diagnosed with stage IV of the disease. "If we had known this earlier in her life, she may have been able to take the appropriate actions to prevent this horrible disease," Cheryl says.

But Sandi's DNA results also resolved her worries about her own lineage. Cheryl and Sandi didn't share half of their DNA, as full siblings do. They shared around 22 percent, making them half-siblings through their mother. Sandi shared 25 percent of her genome with Milton, which meant they were half-siblings through their father, Joe.

Sandi's first reaction was acceptance. Four days after the results came back, she sent Cheryl an e-mail: "The bottom line is that it really doesn't matter. I love you and they loved you . . . in their own way I guess. Daddy was OUR daddy, nothing can change that."

But as the information sunk in, Sandi became distressed about the implications of the test. She was hurt when, a few weeks after the results came through, Cheryl sent an e-mail to Sandi and fifteen other friends and family with the subject line "Who is my biological daddy?"

In the e-mail, Cheryl laid out the whole story of Milton and

Sandi's tests, revealing that Joe wasn't her biological father and reflecting a bit on her feelings:

> My thoughts . . . Daddy Joe knew in his spirit I wasn't his. After all, animals know their child from someone else's, it is part of nature. AND male animals will usually destroy the offspring of other males. And momma never cared for me like she cared for Sandi, why? Because she had a problem with my father, whoever he was. Was she raped, was she involved with someone and he dumped her, or was she just ashamed for some reason. We do not know, as no one is here that we can ask.

Two days later Sandi responded to the e-mail and similar things Cheryl had posted on Facebook. She was upset that Cheryl referred to Joe as Daddy Joe, as opposed to just Daddy. "I know you mean no harm and [are] only trying to distinguish between sperm donor and daddy, but it really bothers me for him," Sandi wrote. "I just feel so bad for him wondering if he knew and now I feel as though he was played a fool."

"I should never have taken that test," she added. "I feel so terribly guilty."

Blaine Bettinger is a well-known figure in the genetic genealogy world. The thirty-seven-year-old is an intellectual-property attorney by day. "But I joke that it's just a way that I make money to pay for more genealogy tests," he says. He has been researching his family history since he was a kid, and he studied molecular biology and genetics in graduate school, so he was perfectly poised for the genetic genealogy revolution. Bettinger bought his first genetic test in 2003. A few years later he launched a blog—The Genetic Genealogist—with the aim of explaining the science behind the tests in simple language. It now receives around a thousand visitors a day, he says.

Genetic genealogy can be extremely complicated, but most cases require only a basic understanding of our twenty-three pairs of chromosomes. Most chromosomes, you could say, are promiscuous. During the formation of egg and sperm, each chromosome inherited from the mother physically crosses with its counterpart from the father, and as the pieces mingle they freely exchange segments of DNA. This recombination gives our species great genetic diversity, and it's the primary reason non-twin siblings are never genetically identical.

But the Y chromosome is chaste. The vast majority of its 50 million DNA letters do not swap with other chromosomes, passing almost identically from father to son, son to grandson, and so on. That means that when a genetic change spontaneously occurs in a Y chromosome, it can be passed down to male descendants forever, serving as a reliable marker of their paternal lineage.

This was famously demonstrated in 1997, when researchers published a study of Jewish priests in the journal *Nature*. According to Jewish belief, the high priesthood began 3,300 years ago with Aaron, Moses's older brother, and has been passed from father to son ever since. Today many Jews have the surname Cohen or Kohen, meaning "priest" in Hebrew. The researchers scraped a few skin cells from the inside of the cheeks of almost two hundred Jewish men from Israel, North America, and England, and compared the men's Y chromosomes. Close to seventy had been told at some point that they were direct descendants of the high priests. And these men, it turns out, had a distinctive Y-chromosome profile. "The simplest, most straightforward explanation is that these men have the Y chromosome of Aaron," the lead researcher told the *New York Times*.

The following year, a similar genetic study made headlines when it bolstered the controversial theory that Thomas Jefferson had fathered a child with his slave, Sally Hemings. The researchers looked at the genes of male descendants of Jefferson's paternal grandfather and found that they carried a combination of nineteen genetic markers that is quite rare, showing up in just a tenth of a percent of all men. But the researchers found exactly the same set of markers in a descendant of one of Hemings's sons, Eston, meaning that Eston's father was either Jefferson or one of Jefferson's close male relatives.

For genealogy buffs, these studies had thrilling implications. Since the Middle Ages, Western cultures have passed surnames from father to son. In theory, then, men who have the same surname should share markers on their Y chromosomes. This wouldn't be true for everyone, of course: multiple families may have taken up the same surname even if they weren't related, and adopted children often take the last name of their adoptive fathers. But it's true for enough people to be useful for tracing family trees.

When Family Tree DNA launched its genetic genealogy test, which screened for twelve markers on the Y chromosome, genealogists could find members of their paternal line not with treks to libraries or cemeteries, but by uploading their DNA results to the company's database.

That was in 2000. By the end of 2001, the company's customers had organized research projects for about a hundred surnames. After 23andMe launched, in 2007, it added thousands of markers associated with health risks, such as those that Cheryl heard about on *Oprah*. There are also companies that specialize in determining ancestry for African Americans, Native Americans, and other specific ethnicities.

For people like Bettinger, DNA testing has made genealogical research richer and more fulfilling. "Once I got the DNA test back, I was able to look at my family tree in a whole new light," he says.

Bettinger is Caucasian and had assumed that his ethnicity was 100 percent European. But tests revealed that he carried Native American markers. "This was a complete shock," he says. He had known from his previous research that some of his ancestors had lived in Honduras in the mid-1800s but had assumed that they were all English missionaries. After getting his results back, he realized that some of them were native Hondurans, with ancestry from both Honduras and the Cayman Islands.

"Genealogy is not only about names, dates, and places, but about filling out the story of each ancestor as well—what their lives were like, what their motivations might have been like, the trials and tribulations and joys that they experienced in their lifetime," Bettinger says. "Every decision, no matter how small, by each one of these individual ancestors ultimately led to my existence—and, undoubtedly, to the person I am today."

For some people who do genetic genealogy, though, the information they unearth is more difficult to accept. "You would not believe the things we can find out," particularly when genetic information is combined with searches from the Internet and social media, says CeCe Moore, the blogger who helped Cheryl. "If you're a privacy advocate, it is worrying."

Over the past three years, Moore says, she has answered e-mails from more than ten thousand people interested in using genetic genealogy and has intensely worked on searches for about a hundred people. Many of these people are adoptees or, like Cheryl, have discovered that they have a mystery father. "We used to only

have an adoptee get a close match every six months," says Moore.
"Now it's happening every single week."

"I believe that knowledge is power, and I think we can gain
much more than we will lose from this movement," says Moore.
(She is also an unpaid liaison between several genetic testing companies and the genetic genealogy community, and a paid consultant for the popular American television show *Finding Your Roots,*
with Henry Louis Gates, Jr.) At the same time, though, there's no
denying that some of this newfound knowledge will be painful. "A
lot of times people find out things that really shake their identity,"
she says.

After learning that Joe Wilmoth wasn't her biological father,
Cheryl began unpacking what she knew about her mother, Vivian.

Vivian Tipton was strikingly beautiful, even in her older years,
and had an infectious cackle of a laugh. She grew up in Petersburg, Virginia, a small town about 25 miles south of Richmond. In
July 1941, when she was sixteen, she married an eighteen-year-old
soldier named Richard Thompson. Just five months later he left to
fight in Europe.

Richard returned after the war, and he and Vivian moved into
a house across from her parents. By the end of 1949, the couple
had two girls, Toni and JoAnn, and Vivian was pregnant with a
third. Vivian always said Richard was the love of her life, but their
marriage was cut short on December 21, 1949, when Richard was
killed in a dump-truck accident. Vivian was devastated, staying in
bed all day and refusing to celebrate Christmas. After having her
third daughter, Jayne, in February, Vivian and the girls moved in
with her parents. Not long after, she moved out, leaving her children to be raised by their grandparents.

The next few years of Vivian's life are not entirely clear, but
sometime in 1950 or 1951 she met Joe Wilmoth. They married in
the summer of 1951 and moved to Chester, about 10 miles from
Petersburg. They had a rocky relationship, to say the least. Joe was
physically abusive at times, and the couple seemed never to stop
arguing. Cheryl was born less than seven months after their marriage, on Christmas Eve of 1951, and grew up believing that she
had been a premature baby. After the DNA test, though, Cheryl
wondered if even that were true. The test raised so many unset-

tling thoughts, the kind that kept her up at night. How many of the other stories of her early life, she wondered, were fiction?

From age one to four, Cheryl lived with a couple who had grown up with Joe. Cheryl doesn't know exactly why. It could be because Joe and Vivian weren't getting along—or, perhaps, because Joe didn't want to raise another man's child.

The next decade of Cheryl's life was unstable and traumatic. She lived in more than a dozen different homes in Florida and Virginia and frequently witnessed violent outbursts from Joe. Looking back, Cheryl suspects that some of Joe's behavior could be explained by posttraumatic stress disorder—he had seen combat in the Philippines during the war. As a child, though, no explanation would have helped. She was only terrified.

Through all of this turmoil, Cheryl tried to protect and care for her little sister, and the girls forged a powerful emotional bond. Still, they were different in more ways than they were alike. Sandi was tall and thin; Cheryl, short and plump. Sandi was happy-go-lucky from a very young age, and by the time she was a young woman, liked to drink, smoke, and party hard. Cheryl was shy, anxious, fearful, and prone to crying.

Perhaps their most striking difference, though, was in their relationship with Joe. "To me, he was everything," Sandi told me. "She was afraid of him."

The day after Christmas 1963, the family moved into a new home. The house was right up the road from Dickie's family, and soon Cheryl and Dickie were sweethearts. She got pregnant in early 1966, soon after her fourteenth birthday, and they were married in May, just before Dickie turned twenty-one.

Cheryl's early marriage shows the extent of Joe and Vivian's parental neglect, Dickie says. "It was pretty obvious," he says. "I mean, you don't let your fourteen-year-old girl go out with somebody as old as I was."

In 1980, when Vivian was just fifty-five, she was diagnosed with lung cancer. Cheryl came over most days to make food and help clean the house. One day she and Vivian opened up an old cedar chest of Vivian's personal mementos. The chest contained a pink card issued by the hospital on the day Cheryl was born. It noted, in handwritten script, that she weighed almost seven pounds— much heavier than a baby who was two months premature could possibly be.

Vivian was sick for three years, and Cheryl's relationship with Joe disintegrated over this time. She has a vivid memory of confronting him one day, when he was sitting at his dining room table. "I said to him, 'Why do you treat me so different from Sandi? What is it? Am I not your child?'"

Joe looked out the window, Cheryl remembers. Then he looked down at his coffee cup and said, "Well, I wouldn't go so far as to say that."

DNA tests, if done rigorously, are far more definitive than tattered forms in old cedar chests, and far more emotionally potent. The genetic genealogy industry had barely gotten off the ground before scientists, sociologists, and ethicists were debating its societal impact—for better and for worse.

Early concerns focused on accuracy. All of the tests—whether they look at the Y chromosome, autosomal chromosomes, or other types of DNA—work in essentially the same way. They screen the billions of letters of a person's DNA for a certain number of markers and then compare that combination of markers with those found in reference samples taken from thousands of people living in various regions of the world. Test accuracy, then, starts with two things: the number of markers analyzed and the size and selection of the comparison set of samples.

When genetic genealogy debuted, the technology cost many times what it does today. So the first tests screened a relatively small number of markers, leading to a crude measure of ancestry. The first test that Blaine Bettinger bought, from a now-defunct company called DNAPrint Genomics, screened his autosomal chromosomes for just seventy-one markers and used those to estimate his ties to four broad ethnic groups: 88 percent "Indo-European," 12 percent "East-Asian," and zero percent "Native-American" and "African."

"Those early autosomal tests were sort of wildly inaccurate," Bettinger says. The subsequent tests he bought showed that his Honduran ancestors had both Native American and African roots.

Today's genetic tests can probe many more markers, making them much more accurate. 23andMe uses around half a million markers, on all twenty-three chromosomes, to probe each customer's ancestry. But many other companies continue to use only a small number of markers, and none make their reference databases or methodology transparent to customers. As one group of scholars wrote in a 2009 paper in *Science:* "Genetic ancestry tests fall into an unregulated no-man's land, with little oversight and few industry guidelines to ensure the quality, validity, and interpretation of information sold."

Even if all the tests were completely accurate, they'd still pose big philosophical questions: How much weight do individuals give to genes when forming ethnic, racial, and religious identities? How much weight should they give to DNA?

Cheryl, when pressed, acknowledges that she is some combination of her innate genetic predisposition, traumatic upbringing, and six decades of life experience. But like many people, she seems to give special weight to her genes. As she posted to a site called Cousin Connect shortly after finding out that Joe wasn't her father: "I want nothing from anyone [except] to know what blood line flows through my veins, my children, and grandchildren's."

Many people find religious and cultural identities in their DNA. Take Andrea, a thirty-five-year-old adoptee. When she was less than two years old, her biological parents put her and two older brothers in foster care. The children were soon adopted, but their new parents were alcoholics, and they had messy and difficult childhoods.

At sixteen Andrea left home and, a few years later, began searching for her birth parents. She found her father's profile on a dating web site and called him. "That was a really hard phone call, because he was not interested in me," she says. He also told her about her biological mother, and some searching revealed that she had died. "It was very upsetting."

But Andrea was profoundly uplifted by the results of her DNA test, which she bought from 23andMe about a year ago. The test indicated that she's approximately one-quarter Ashkenazi Jewish. "That was like, the shock of all shocks," she says. Though she is a practicing Christian, she has felt strong ties to Jewish culture since college, where she was a religious studies major. "I was very, very drawn to Jewish studies classes, I took biblical Hebrew, and always wanted to go to Israel," she says.

Finding out that she had genetic roots to Judaism was bittersweet, she says, because she would have liked to have grown up in the Jewish culture. She's making up for it now by reading all she can about Jewish history. "I will sit at home and watch documentaries on YouTube about Jerusalem," she says, laughing. "I love it. And it's so fascinating to me—the personal connection I had [with Judaism] even before I knew, and the one that continues now in my life."

That kind of emotional connection, the "Aha!" moment, is what Cheryl has been searching for all her life. She's always wondered why she and her children don't look like her sisters, Vivian, or Joe. Her son, Travis Whittle, has curly hair and a gregarious personality that Cheryl says resemble no one else in the family.

She wonders about her own traits and predispositions too. She has had several bouts of depression over the years and is almost always anxious. "I know that my mother had some depression. But I wonder if my father might have had some problems too."

It wasn't enough to know that Joe wasn't her father. To feel whole, she had to know who her real father was. "You know how in the Bible it says so-and-so begot so-and-so begot so-and-so?" Cheryl says. "If you leave out a begot, there's something missing. It doesn't quite fit."

Cheryl's fervent hunt for the mystery man responsible for half of her genetic identity has consumed much of her time over the past three years.

She didn't have much luck browsing the Relative Finder section of the 23andMe web site, which compared her DNA to that of the other people in the company's database who had opted to share with the community. Her only genetic matches were estimated fourth, fifth, and sixth cousins—nowhere near close enough to trace back to her father.

In June 2010, Cheryl bought Family Tree DNA's genetic test, "to fish in more ponds," as she puts it. The $293 cheek-swab test gave her access to all of the people in Family Tree DNA's database, any one of which could have been a match. Unfortunately, though, she caught no fish.

Cheryl's search went cold for nearly two years before picking up in April 2012, when CeCe Moore put her in touch with Diane Harman-Hoog of Redmond, Washington. Diane has spent her retirement years—"seventeen hours a day, seven days a week," she says—as a genealogy "search angel," helping hundreds of people, mostly adoptees, figure out their family mysteries at no charge. Diane had just started to add genetic results into her search methods and was eager to look into Cheryl's case.

By pooling information from various sites, Diane created a spreadsheet showing more than five hundred people who shared some of Cheryl's DNA. Each line of the spreadsheet gave the person's surname and the precise chromosome location where their DNA matched Cheryl's. But, big as it was, Diane's spreadsheet didn't identify any useful leads. "Diane wrote me and said, 'Cheryl, we do not have enough. Your matches aren't close enough yet,'" Cheryl recalls.

That was in August 2012. Just a few months earlier, Ancestry .com—the largest genealogy company in the world—began selling its first autosomal DNA test. Cheryl bought one in September for $99, to try her luck in yet another pond. Eight weeks later she had her matches: nobody was closer than a fourth cousin.

During this lull in her search, Cheryl says, she had a profound spiritual experience. "One morning I got out of bed—and this sounds crazy probably, but, you know, I believe in God. I was just feeling real down about it. But then something inside of me said, 'You will find your father.' And so I was clinging to that. I knew that it's just a matter of not losing hope."

One February morning this year, Cheryl received a note via Ancestry.com's internal message service from a woman named Jeannette Morrison, a genealogy hobbyist who had taken the test to expand her tree. She had identified Cheryl as a possible second cousin. Cheryl wrote back immediately and updated Diane about the new lead. A second-cousin match, Cheryl knew, could be a very big deal.

While Ancestry.com's test will estimate the relatedness of two people, it doesn't allow customers to compare their genetic data chromosome by chromosome. And Cheryl couldn't tell from Jeannette's family tree whether they were related through Cheryl's mother or father. So Cheryl bought another 23andMe kit and sent it to Jeannette's house in Ohio.

When Jeannette's test results came back, in April, Cheryl discovered that Jeannette did not share any DNA with Sandi. In other words, Jeannette was exactly what Cheryl had been praying for: a solid lead to her biological father.

23andMe showed that Jeannette and Cheryl shared seventeen

segments of their DNA—including, crucially, two bits of the X chromosome. Through logical inferences and painstaking searches —comparing trees, geographical locations, birth and death dates —Diane found one of Jeannette's relatives, Joseph Parker, who was about the right age and had lived in the right place to be Cheryl's biological father.

Joseph had died in 1987, leaving behind a son, Joseph Jr., and one daughter, Effie Jane, who lived in Richmond. According to Diane's analysis, Effie Jane could be either Cheryl's first cousin once removed or her half-sister.

There was only one way to find out.

Genetic genealogy is part of the much broader cultural trend of uploading personal data to the cloud. We willingly flaunt photos, videos, and demographic information on social media—Facebook, Twitter, Flickr, match.com—and give our credit card and social security numbers to banking and retail sites. Even seemingly private data—e-mails, cell phone records, Internet browsing patterns—is actually, we're learning, under government surveillance.

What's fascinating about genetic genealogy is that it brings together two very different perspectives on privacy. DNA is arguably as personal as it gets. It's an individual's unique code of life. That's why, among doctors and health care workers, genetic data is subject to strict privacy regulations.

The traditional genealogy community, on the other hand, is all about sharing—sharing family trees, sharing documents, sharing stories. "The only way you can connect with people is with some loss of privacy," says Yaniv Erlich, a geneticist at the Whitehead Institute for Biomedical Research in Cambridge, Massachusetts.

Erlich is a world expert on genetic privacy. Earlier this year his team caused a stir among medical researchers with a study in *Science* showing that supposedly anonymous participants in genetic research studies can be identified using simple software and an Internet hookup.

But Erlich is also an avid genealogist. In the past few years he has bought DNA tests from 23andMe and Family Tree DNA and has chosen to upload his genetic data to their databases. In doing so, he discovered that he carries the Cohen profile on his Y chromosome, confirming what had been passed down through his family's oral tradition. He also found a fifth cousin, whom he later met at a family reunion in Poland. That cousin grew up as a Christian, but because of his genetic discoveries is converting to Judaism. "It touches people, what they find in their DNA," Erlich says. "I think it's wonderful."

The core privacy tension in genetic genealogy, Erlich notes, is that your DNA is not yours alone. "By putting your data out there," Erlich says, "you're not only sacrificing your own privacy but also the privacy of people who are connected to you, because you share DNA."

In June the *Times,* a British newspaper, ran a front-page story with the headline "Revealed: The Indian Ancestry of William." Two distant cousins of Prince William had their DNA tested with a company called BritainsDNA and discovered that they carried a rare set of markers that had previously been found in only fourteen people: thirteen Indian and one Nepalese. Because the DNA in question passes only from mother to child, and the cousins shared a great-great-great-great-grandmother with William's mother, Diana, they could infer that the heir to the throne also has these Indian roots.

That particular bit of trivia is only important if you're in the business of selling newspapers, as commentator Alex Hern pointed out in the *New Statesman.* But what if the genetic intel hadn't been so silly? As Hern put it, "There is a wider issue at stake here, which is that the story reveals information about the genetic makeup of someone who has not consented to any DNA tests."

The loose definition of genetic privacy, of course, is what allows people like Cheryl to solve their life mysteries. Cheryl's cousin Jeannette, by agreeing to a DNA test, opened up the possibility of Cheryl identifying her real father and his descendants—regardless of whether any of them wanted segments of their DNA posted on a public database.

This risk—that relatives may be harmed in some way by the sharing of their DNA—has led some to argue that the decision to share is not an individual's to make. In 2010 Henry Louis Gates Jr. asked twelve celebrities to get DNA tests for his television show about genealogy, *Faces of America.* The novelist Louise Erdrich was the only one to refuse. Erdrich's maternal grandfather was a chief of the Turtle Mountain Band of Chippewa Indians, a Native American tribe in North Dakota, and Erdrich is also an enrolled member. As Erdrich explained to Gates regarding the DNA test: "It wouldn't do me any harm, but when I asked my extended family about this—and I did go to everyone—I was told, 'It's not yours to give, Louise.'"

Legally, however, genetic testing is an individual decision. And unlike Erdrich, many people only consider privacy repercussions when they're suddenly facing them.

Mike Taffe hadn't given much thought to violating his extended family's privacy when he sent a tube of his spit to 23andMe in early 2012. Taffe is a neuroscientist at the Scripps Research Institute in La Jolla, California, and he was interested in the medical risk markers he might be carrying. As an adoptee, he would also become interested in the company's Relative Finder service.

For Taffe's first few months on the service, the closest matches were not close: third or fourth cousins. Then a first cousin popped up: an African American man named Chris. Taffe wrote to Chris using 23andMe's messaging service, explaining that he was looking for his father, who was, according to records from the adoption agency, a Puerto Rican man born in the 1940s. Chris said that didn't ring any bells, and their correspondence ended soon after.

Taffe let it go. Then, a year later, the company dropped its prices to $99 per kit, spurring him to send in spit samples of his three children.

Back on the site, he was reminded of his message to Chris and started snooping around online. He found Chris's Facebook page, which was open to public viewing, and clicked through his photos. From these he spotted Chris's mother. (They had to be related through her, Taffe reasoned, because 23andMe had shown that they didn't share Y-chromosome DNA.) Then, on the Facebook page of his presumed aunt, Taffe found some of her high school photos and her name.

After some more judicious Googling, he found an obituary for his aunt's brother. No other siblings were mentioned. This man, Taffe thought, could very well be his biological father. The obit named the man's three surviving children, and one of them, Cliff, had a Facebook page.

Taffe was immediately struck by Cliff's photo: they shared a nose. "It was like, whoa, dude," Taffe says. Looking through his presumed half-brother's photos, Taffe saw a few that Cliff had posted of his father around the time of his death. The nose was the same.

This summer Taffe took the plunge and sent a Facebook message to Cliff. It was vague, saying only that he was interested in ancestry and that he thought they might be related. Cliff never responded. If he ever does, Taffe isn't sure how much more he would disclose. "People have a right to privacy, even from their relatives," Taffe says.

He has started a discussion about these issues on a 23andMe forum targeted at adoptees. "Obviously a lot of the adoptees have decided that their right to know trumps any other possible consideration. I do not, at present, agree with this."

The 23andMe community forums, open to any 23andMe customer, are filled with gripping stories of people who are looking for their parents. They're usually adoptees who bought the 23andMe test for this purpose. But there are also stories of people who inadvertently discovered that their parents weren't who they said they were.

One such story comes from Terry, a college student whose world was rocked by the matches she found on 23andMe. At the top of the list was a man whom she calls John Doe, who shares 28 percent of his DNA with her. 23andMe suggested that he was her grandfather, but that didn't make sense, given what she knew about her family.

Terry sent John Doe a message through 23andMe and learned that he is twenty-three years old, ruling out the grandfather theory. After hunting around online, she discovered that it's not only her grandparents who share one-quarter of her DNA; John Doe could also be her uncle, half-brother, or double first cousin. Since she didn't share any X-chromosome pieces with John Doe, she figured he was a half-brother through her father.

Terry called her mother and asked her what she made of John Doe. As Terry explained on the forum, after initially seeming baffled, her mother tried to deter her from pursuing John Doe. Over the next few days, dozens of other community members chimed in with advice and sympathy. One member asked Terry if she had considered the possibility that her parents had used a sperm donor. She had. Her parents were married for nine years before her birth, she wrote, so it was possible they had had fertility issues. After learning about John Doe, Terry had asked her mother directly if they had used a sperm donor—a suggestion her mother described as "insulting." Terry believed her mother at first. But doubts crept in, partly because Terry's ancestry composition did not match her father's ethnicity.

John Doe, for his part, didn't seem to understand genetics. He told Terry he thought they were cousins, and that it was just a "co-incidence" that they shared so much DNA. Terry gently reminded him that first cousins share 12.5 percent of their DNA on average —not 28 percent.

Eventually, Terry's parents confessed: they had indeed used a sperm donor. Despite the new information, John Doe was in de-nial, continuing to say that he was definitely his father's son.

Terry's story illustrates the web of privacy concerns that these tests can create. Her results challenged the privacy of her parents, her half-brother, and her half-brother's parents. Terry's results may ultimately affect the privacy of her biological father as well. She has sent messages to several people the database identified as her first cousins, thinking that they might know an uncle who donated sperm.

Terry would have preferred that her parents had told her the truth, but she understands why they didn't. In fact, she's now keep-ing the same secret from a younger sibling who was conceived with a different sperm donor. As someone on the forum pointed out, Terry will have to keep her sibling away from 23andMe.

23andMe spells out several privacy issues in its Terms of Service agreement. In the fifth of twenty-eight sections, titled "Risks and Considerations Regarding 23andMe Services," the company states several stark facts in bold type, including:

- **Once you obtain your Genetic Information, the knowledge is irre-vocable.**
- **You may learn information about yourself that you do not anticipate.**
- **Genetic Information you share with others could be used against your interests.**

In a different document, called the Privacy Statement, the com-pany highlights possible consequences of sharing genetic informa-tion. "Personal Information, once released or shared, can be diffi-cult to contain," it reads. "It is incumbent upon customers to share Personal Information only with people they know and trust."

Family Tree DNA and Ancestry.com have similar agreements, though they're less explicit about these intangible risks. It's impossible to know, of course, how many customers ever read them. "We'd like to believe that everyone who's accepted the terms has actually read them, but in practice we know that that's not always the case," says Catherine Afarian, a spokesperson for 23andMe.

As the company's database grows, there are bound to be more people who find out family secrets. "It's a reflection of what actually happens in our society," Afarian says. Privacy is a top concern for the company, she adds, which is why the consent forms and privacy documents are written in clear and explicit language. "We try to be really up front." For now, these issues haven't caused any legal problems or customer upsets, she says, and if they do, the company's privacy policy "could certainly evolve over time."

Nor is privacy the only area where DNA testing's ethics are considered a work in progress. In November, 23andMe came under fire from the U.S. Food and Drug Administration for the part of its service that offers medical risk profiles. Customer results, delivered online with brief explanations of genetic indicators, such as Sandi's discovery that she was at high risk of cancer, therefore fall under the FDA's regulatory regime.

The agency accused the company of failing to cooperate fully over a period of four years and of failing to communicate at all for more than six months. And although 23andMe rebuffed the claims, the company has now stopped marketing its services and faces a class-action lawsuit from disgruntled customers.

Cheryl has bought sixteen kits from 23andMe, two for her (after her first one, she bought an updated version) and fourteen on behalf of her friends and family. For each of them, she asked their permission and explained what she wanted to use the data for. But nobody ever read the forms.

Still, not everybody is interested in playing the genetic genealogy game. When Cheryl asked her daughter Wendy's husband, Dennis Plear, if he'd like to test, he flatly refused and also forbade Wendy from having their children tested. Dennis, a disabled navy veteran who is half African American and half German, is wary of such personal information winding up in the hands of the government. "When you start giving out samples of your DNA, you're opening the door for other people to be in your business," he says.

When told that these companies purport to be keeping the information secure and out of reach of third parties, Dennis wasn't impressed. "If you have it on file, on the Internet, the government has access to it one way or another."

But for every customer deterred by privacy concerns, there seem to be many more who don't care. Genetic genealogy databases are growing every day, and that pace will quicken as the costs of genetic sequencing drop.

It's plausible that in the not too distant future, we'll all be identifiable in genetic databases, whether through our personal contribution or that of our relatives. Some ethicists have called for federal legislation to restrict these kinds of collections. But many scientists, including Erlich, are against that idea, pointing out that genetic databases can be used for good—not only for genealogy, but for medical discoveries. "We cannot change the course of technology," Erlich says.

Instead, he says, regulation should focus on preventing harm to individuals. For example, the Genetic Information Nondiscrimination Act (GINA), passed in the United States in 2008, says that health insurers and employers cannot use an individual's genetic information to deny medical coverage or to make employment decisions. "GINA was a good step forward," Erlich says. "It's not about having the information—it's about not abusing it."

From her training as a nurse, Cheryl is aware of the need to respect other people's privacy. So in May of this year, when she first reached out to Effie Jane, she was wary of divulging too much, saying vaguely that she was interested in the Parker family line.

About a week later Effie Jane wrote her back saying she would be happy to talk to her. Over the next couple of weeks, Cheryl and Effie Jane shared many e-mails and phone calls and discovered that they had a lot in common.

"I never felt like I was a part of my family," Effie Jane says. Her ruddy complexion and crystal blue eyes look like Cheryl's, but that's the extent of their physical resemblance. Their childhoods, though, were similarly traumatic. Effie Jane started having seizures at six months old, and at fourteen her mother died. Her father, Joseph Parker, like Cheryl's adoptive father, Joe, was cold and indifferent, and "may have had a little mental imbalance," Effie Jane says. After her mother died, his personality changed; for example, he instructed her to start calling him "Mr. Parker" rather than "Daddy."

Effie Jane has felt rejected her whole life, but found acceptance in Cheryl. "I had always prayed for a sister," she says. After meeting for the first time at Panera, the women kept talking and slowly began to broach the sensitive subject on both of their minds: If they were indeed sisters, then how would Vivian have known Mr. Parker?

Mr. Parker was a train engineer and would often be away from home for days at a time. Vivian frequently took the same train he worked on to vacation with her family. So she could have known him for years. Or perhaps she met him just once. Maybe they were friendly, or, who knows, maybe she was raped. These were the theories that Cheryl and Effie Jane tossed around. They wouldn't know for sure until the test results came back. But Cheryl felt in her gut that this was the one.

"I'm feeling really comfortable with where I'm going with this. I feel like I'm going to find answers soon." That's what Cheryl said to me during our first phone interview, in early July. She had been anxiously tracking the shipment of Effie Jane's kit and knew that it had arrived at 23andMe that Saturday. She had just a few weeks to wait.

Cheryl and Effie Jane's relationship blossomed over that period. As Cheryl described in an e-mail to Sandi, "There are a lot of things about Effie that seem familiar to me, as though we are of the same cloth."

Cheryl and Effie Jane went to the cemetery where Mr. Parker is buried and snapped photos of his plaque. They also attended an all-day church workshop put on by Cheryl's son, Travis, about how to deal with rejection. "My mom has a hard time believing that we accept her," Travis says. "But that's because of the deck of cards that she's playing with."

On July 23 Cheryl wrote to me with the surprising verdict: Effie Jane was not her sister. She was not even a first cousin once removed, as Diane had guessed. The genetic test estimated them as fourth cousins, at most, with just two shared segments of chromosome.

I asked Cheryl how she was feeling about it, and her response read, to me, like she was trying to hide her disappointment: "I am good! I am frustrated, as now we will have to find someone else to test, and wait yet again!"

Effie Jane took the news in stride. "I didn't cry. I didn't have that feeling of being deserted by her," she told me. And the fact that you don't share much DNA, does that change how you feel about your relationship, I asked. "As far as I'm concerned, she's my little sister."

Cheryl, too, says she will probably be lifelong friends with Effie Jane. But there was still a hole in her personal history. And it needed to be filled.

Effie Jane's test results weren't what Cheryl was hoping for, but they did help her and Diane to focus on the relevant part of Jeannette's family tree. Diane quickly identified another man as Cheryl's potential father, and Cheryl tracked down the phone number of the man's daughter, Rose.

Cheryl called Rose one morning in late July. The conversation went surprisingly well. "She said, 'Oh, are you saying you might be my sister? Oh, I hope you might be my sister!'" Cheryl recalls. Rose agreed to take a 23andMe test, and they tentatively planned to get together the following day to do the spitting.

Late that evening, Cheryl had an upsetting phone call with Rose. The DNA test had been weighing on her the entire day, Rose said, and she had decided she didn't want to tarnish the memory of her father by taking it. "Of course I was in tears," Cheryl told me the next morning. "It's like another rejection."

Cheryl says she understands how Rose must be feeling. "I am nobody to them." She is now trying to decide whether she should reach out to Rose's late brother's wife, to ask if her children might get tested. Cheryl knows that Rose would be upset if she did that; on the other hand, there may be no other way to find out who her father is. "It will make me feel more complete to know who I came from."

Though she doesn't fully realize it, Cheryl is playing out the hypothetical scenario painted by bioethicists: How does one person's right to know stack up against another's right to privacy?

While she struggles with her search for a new family, Cheryl is trying to mend the tears that her genetic testing has created among the relatives she already has.

In June of last year the *Village News,* the local newspaper in
Chester, ran a story about Cheryl's search for her biological father.
The article included several factual inaccuracies, as well as a wildly
inaccurate description of Joe as a man who "stepped up to the
plate making Vivian, as they said then, an honest woman."

All of Cheryl's sisters were upset by the article. Sandi was upset
because of what it implied about Vivian. "[It] shed a somewhat
dim light of my mom and some of the report was inaccurate and
fabricated," she told me. "I only want my parents' legacy to be respected."

On the other hand, Cheryl's older siblings, her half-sisters by
Vivian's first husband, were crushed by the description of Joe as
an upstanding husband and father. "My sisters were furious. They
really wanted to go for blood," says one of them, JoAnn Lear,
who, unlike Cheryl, is tall and boisterous, with a tell-it-like-it-is de-
meanor. They thought about suing the newspaper but decided the
damage had already been done.

I met JoAnn this past August, during a weekend I spent with
Cheryl and Dickie at their home. On Saturday afternoon Cheryl
organized a family reunion of sorts, so that I could meet her children, grandchildren, and Effie Jane. By the time I arrived, I had
been talking to Cheryl on and off for several weeks about her hunt
for her biological father. I was expecting to observe strained or
awkward family relationships—holes that Cheryl was trying to fill
with her genetic search. Instead I was struck by the love, humor,
and openness of Cheryl's family, new and old. When Effie Jane
—Cheryl's newest fourth cousin—arrived, she hugged Cheryl and
gave her a gift: a sand dollar. It was a "Holy Ghost shell," she explained, with the star in the center representing the Star of Beth-
lehem.

Sandi wasn't at the reunion—her colorectal cancer had by then
spread to her lungs, and it was uncomfortable for her to leave her
house or even to speak on the phone. (All of our discussions for
this article happened through Facebook chats and e-mails; Sandi
died on September 6, 2013.) Cheryl's three older sisters weren't
there, either. None of them had talked to Cheryl much since the
infamous *Village News* article had come out.

JoAnn, though, had agreed to meet with us the next day. So
Cheryl and I made the pretty, winding drive from her riverfront
home to the suburbs of Richmond, where JoAnn lives, and picked

her up for lunch and a ride around the small towns where they grew up. Cheryl and JoAnn had lost touch in recent years, but had recently reunited to care for Sandi.

After about an hour and a half of driving around, we pulled into a quiet road called Mason Avenue in Chester. "It was one of these two," said Cheryl from the back seat, pointing to two small houses. Each house used to hold two apartments, and one of them is where Cheryl, Sandi, Joe, and Vivian once lived.

This is the place, Cheryl said, where Joe once pulled a gun on her and Sandi and threatened to kill them.

"We were there," JoAnn said.

"You were?" Cheryl said, astonished. "I wonder if it was the same incident."

"He was drunk. And he said he was going to kill us all," JoAnn continued, her voice welling up with emotion. "We got out of the house and we took off running, and he was shooting towards us."

JoAnn continued with her memory of the story, and then Cheryl recounted hers. Then the car got quiet.

After a minute, Cheryl finally said, "You know, JoAnn, we really do need to talk these things out between us."

"I know, I know," JoAnn said.

FERRIS JABR

Why the Brain Prefers Paper

FROM *Scientific American*

ONE OF THE MOST PROVOCATIVE viral YouTube videos in the past two years begins mundanely enough: a one-year-old girl plays with an iPad, sweeping her fingers across its touch screen and shuffling groups of icons. In following scenes, she appears to pinch, swipe, and prod the pages of paper magazines as though they, too, are screens. Melodramatically, the video replays these gestures in close-up.

For the girl's father, the video—*A Magazine Is an iPad That Does Not Work*—is evidence of a generational transition. In an accompanying description, he writes, "Magazines are now useless and impossible to understand, for digital natives"—that is, for people who have been interacting with digital technologies from a very early age, surrounded not only by paper books and magazines but also by smartphones, Kindles, and iPads.

Whether or not his daughter truly expected the magazines to behave like an iPad, the video brings into focus a question that is relevant to far more than the youngest among us: How exactly does the technology we use to read change the way we read?

Since at least the 1980s researchers in psychology, computer engineering, and library and information science have published more than one hundred studies exploring differences in how people read on paper and on screens. Before 1992 most experiments concluded that people read stories and articles on screens more slowly and remember less about them. As the resolution of screens on all kinds of devices sharpened, however, a more mixed set of findings began to emerge. Recent surveys suggest that although most people still prefer paper—especially when they need to concentrate for a long time—attitudes are changing as tablets and e-reading technology improve and as reading digital texts for facts and fun becomes more common. In the United States, e-books currently make up more than 20 percent of all books sold to the general public.

Despite all the increasingly user-friendly and popular technology, most studies published since the early 1990s confirm earlier conclusions: paper still has advantages over screens as a reading medium. Together, laboratory experiments, polls, and consumer reports indicate that digital devices prevent people from efficiently navigating long texts, which may subtly inhibit reading comprehension. Compared with paper, screens may also drain more of our mental resources while we are reading and make it a little harder to remember what we read when we are done. Whether they realize it or not, people often approach computers and tablets with a state of mind less conducive to learning than the one they bring to paper. And e-readers fail to re-create certain tactile experiences of reading on paper, the absence of which some find unsettling.

"There is physicality in reading," says the cognitive scientist Maryanne Wolf of Tufts University, "maybe even more than we want to think about as we lurch into digital reading—as we move forward perhaps with too little reflection. I would like to preserve the absolute best of older forms but know when to use the new."

Textual Landscapes

Understanding how reading on paper differs from reading on screens requires some explanation of how the human brain interprets written language. Although letters and words are symbols representing sounds and ideas, the brain also regards them as physical objects. As Wolf explains in her 2007 book *Proust and the Squid,* we are not born with brain circuits dedicated to reading, because we did not invent writing until relatively recently in our evolutionary history, around the fourth millennium B.C. So in childhood the brain improvises a brand-new circuit for reading by weaving together various ribbons of neural tissue devoted to other abilities, such as speaking, motor coordination, and vision.

Some of these repurposed brain regions specialize in object recognition: they help us instantly distinguish an apple from an orange, for example, based on their distinct features, yet classify both as fruit. Similarly, when we learn to read and write, we begin to recognize letters by their particular arrangements of lines, curves, and hollow spaces—a tactile learning process that requires both our eyes and our hands. In recent research by Karin James of Indiana University Bloomington, the reading circuits of five-year-old children crackled with activity when they practiced writing letters by hand but not when they typed letters on a keyboard. And when people read cursive writing or intricate characters such as Japanese *kanji*, the brain literally goes through the motions of writing, even if the hands are empty.

Beyond treating individual letters as physical objects, the human brain may also perceive a text in its entirety as a kind of physical landscape. When we read, we construct a mental representation of the text. The exact nature of such representations remains unclear, but some researchers think they are similar to the mental maps we create of terrain—such as mountains and trails—and of indoor physical spaces, such as apartments and offices. Both anecdotally and in published studies, people report that when trying to locate a particular passage in a book, they often remember where in the text it appeared. Much as we might recall that we passed the red farmhouse near the start of a hiking trail before we started climbing uphill through the forest, we remember that we read about Mr. Darcy rebuffing Elizabeth Bennet at a dance on the bottom left corner of the left-hand page in one of the earlier chapters of Jane Austen's *Pride and Prejudice.*

In most cases, paper books have more obvious topography than on-screen text. An open paper book presents a reader with two clearly defined domains—the left- and right-hand pages—and a total of eight corners with which to orient oneself. You can focus on a single page of a paper book without losing awareness of the whole text. You can even feel the thickness of the pages you have read in one hand and the pages you have yet to read in the other. Turning the pages of a paper book is like leaving one footprint after another on a trail—there is a rhythm to it and a visible record of how far one has traveled. All these features not only make the text in a paper book easily navigable, they also make it easier to form a coherent mental map of that text.

In contrast, most digital devices interfere with intuitive navigation of a text and inhibit people from mapping the journey in their mind. A reader of digital text might scroll through a seamless stream of words, tap forward one page at a time, or use the search function to immediately locate a particular phrase—but it is difficult to see any one passage in the context of the entire text. As an analogy, imagine if Google Maps allowed people to navigate street by individual street, as well as teleport to any specific address, but prevented them from zooming out to see a neighborhood, state, or country. Likewise, glancing at a progress bar gives a far more vague sense of place than feeling the weight of read and unread pages. And although e-readers and tablets replicate pagination, the displayed pages are ephemeral. Once read, those pages vanish. Instead of hiking the trail yourself, you watch the trees, rocks, and moss pass by in flashes, with no tangible trace of what came before and no easy way to see what lies ahead.

"The implicit feel of where you are in a physical book turns out to be more important than we realized," says Abigail J. Sellen of Microsoft Research Cambridge in England, who coauthored the 2001 book *The Myth of the Paperless Office*. "Only when you get an e-book do you start to miss it. I don't think e-book manufacturers have thought enough about how you might visualize where you are in a book."

Exhaustive Reading

At least a few studies suggest that screens sometimes impair comprehension precisely because they distort people's sense of place in a text. In a January 2013 study by Anne Mangen of the University of Stavanger in Norway and her colleagues, seventy-two tenth-grade students studied one narrative and one expository text. Half the students read on paper, and half read PDF files on computers. Afterward, students completed reading comprehension tests, during which they had access to the texts. Students who read the texts on computers performed a little worse, most likely because they had to scroll or click through the PDFs one section at a time, whereas students reading on paper held the entire texts in their hands and quickly switched between different pages. "The ease with which you can find out the beginning, end, and everything in between and the constant connection to your path, your progress in the text, might be some way of making it less taxing cognitively," Mangen says. "You have more free capacity for comprehension."

Other researchers agree that screen-based reading can dull comprehension because it is more mentally taxing and even physically tiring than reading on paper. E-ink reflects ambient light just like the ink on a paper book, but computer screens, smartphones, and tablets shine light directly on people's faces. Today's LCDs are certainly gentler on eyes than their predecessor, cathode-ray tube (CRT) screens, but prolonged reading on glossy, self-illuminated screens can cause eyestrain, headaches, and blurred vision. In an experiment by Erik Wästlund, then at Karlstad University in Sweden, people who took a reading comprehension test on a computer scored lower and reported higher levels of stress and tiredness than people who completed it on paper.

In a related set of Wästlund's experiments, eighty-two volunteers completed the same reading comprehension test on computers, either as a paginated document or as a continuous piece of text. Afterward, researchers assessed the students' attention and working memory—a collection of mental talents allowing people to temporarily store and manipulate information in their mind. Volunteers had to quickly close a series of pop-up windows, for example, or remember digits that flashed on a screen. Like many cognitive abilities, working memory is a finite resource that diminishes with exertion.

Although people in both groups performed equally well, those who had to scroll through the unbroken text did worse on the attention and working-memory tests. Wästlund thinks that scrolling—which requires readers to consciously focus on both the text and how they are moving it—drains more mental resources than turning or clicking a page, which are simpler and more automatic gestures. The more attention is diverted to moving through a text, the less is available for understanding it. A 2004 study conducted at the University of Central Florida reached similar conclusions.

An emerging collection of studies emphasizes that in addition to screens possibly leaching more attention than paper, people do not always bring as much mental effort to screens in the first place. Based on a detailed 2005 survey of 113 people in northern California, Ziming Liu of San Jose State University concluded that those reading on screens take a lot of shortcuts—they spend more time browsing, scanning, and hunting for keywords compared with people reading on paper and are more likely to read a document once and only once.

When reading on screens, individuals seem less inclined to engage in what psychologists call metacognitive learning regulation —setting specific goals, rereading difficult sections, and checking how much one has understood along the way. In a 2011 experiment at the Technion–Israel Institute of Technology, college students took multiple-choice exams about expository texts either on computers or on paper. Researchers limited half the volunteers to a meager seven minutes of study time; the other half could review the text for as long as they liked. When under pressure to read quickly, students using computers and paper performed equally well. When managing their own study time, however, volunteers using paper scored about 10 percentage points higher. Presumably, students using paper approached the exam with a more studious attitude than their screen-reading peers and more effectively directed their attention and working memory.

Even when studies find few differences in reading comprehension between screens and paper, screen readers may not remember a text as thoroughly in the long run. In a 2003 study Kate Garland, then at the University of Leicester in England, and her team asked fifty British college students to read documents from an introductory economics course either on a computer monitor or in a spiral-bound booklet. After twenty minutes of reading, Garland and her colleagues quizzed the students. Participants scored equally well regardless of the medium but differed in how they remembered the information.

Psychologists distinguish between remembering something—a relatively weak form of memory in which someone recalls a piece of information, along with contextual details, such as where and when one learned it—and knowing something: a stronger form of memory defined as certainty that something is true. While taking the quiz, Garland's volunteers marked both their answer and whether they "remembered" or "knew" the answer. Students who had read study material on a screen relied much more on remembering than on knowing, whereas students who read on paper depended equally on the two forms of memory. Garland and her colleagues think that students who read on paper learned the study material more thoroughly more quickly; they did not have to spend a lot of time searching their mind for information from the text—they often just knew the answers.

Perhaps any discrepancies in reading comprehension between paper and screens will shrink as people's attitudes continue to change. Maybe the star of *A Magazine Is an iPad That Does Not Work* will grow up without the subtle bias against screens that seems to lurk among older generations. The latest research suggests, however, that substituting screens for paper at an early age has disadvantages that we should not write off so easily. A 2012 study at the Joan Ganz Cooney Center in New York City recruited thirty-two pairs of parents and three- to six-year-old children. Kids remembered more details from stories they read on paper than ones they read in e-books enhanced with interactive animations, videos, and games. These bells and whistles deflected attention away from the narrative toward the device itself. In a follow-up survey of 1,226 parents, the majority reported that they and their children prefer print books over e-books when reading together.

Nearly identical results followed two studies, described this past September in *Mind, Brain, and Education,* by Julia Parrish-Morris, now at the University of Pennsylvania, and her colleagues. When reading paper books to their three- and five-year-old children, parents helpfully related the story to their child's life. But when reading a then popular electric console book with sound effects, parents frequently had to interrupt their usual "dialogic reading" to stop the child from fiddling with buttons and losing track of the narrative. Such distractions ultimately prevented the three-year-olds from understanding even the gist of the stories, but all the children followed the stories in paper books just fine.

Such preliminary research on early readers underscores a quality of paper that may be its greatest strength as a reading medium: its modesty. Admittedly, digital texts offer clear advantages in many different situations. When one is researching under deadline, the convenience of quickly accessing hundreds of keyword-searchable online documents vastly outweighs the benefits in comprehension and retention that come with dutifully locating and rifling through paper books one at a time in a library. And for people with poor vision, adjustable font size and the sharp contrast of an LCD screen are godsends. Yet paper, unlike screens, rarely calls attention to itself or shifts focus away from the text. Because of its simplicity, paper is "a still point, an anchor for the consciousness," as William

Powers writes in his 2006 essay "Hamlet's Blackberry: Why Paper Is Eternal." People consistently report that when they really want to focus on a text, they read it on paper. In a 2011 survey of graduate students at National Taiwan University, the majority reported browsing a few paragraphs of an item online before printing out the whole text for more in-depth reading. And in a 2003 survey at the National Autonomous University of Mexico, nearly 80 percent of 687 students preferred to read text on paper rather than on a screen to "understand it with clarity."

Beyond pragmatic considerations, the way we feel about a paper book or an e-reader—and the way it feels in our hands—also determines whether we buy a best-selling book in hardcover at a local bookstore or download it from Amazon. Surveys and consumer reports suggest that the sensory aspects of reading on paper matter to people more than one might assume: the feel of paper and ink; the option to smooth or fold a page with one's fingers; the distinctive sound a page makes when turned. So far digital texts have not satisfyingly replicated such sensations. Paper books also have an immediately discernible size, shape, and weight. We might refer to a hardcover edition of Leo Tolstoy's *War and Peace* as a "hefty tome" or to a paperback of Joseph Conrad's *Heart of Darkness* as a "slim volume." In contrast, although a digital text has a length that may be represented with a scroll or progress bar, it has no obvious shape or thickness. An e-reader always weighs the same, regardless of whether you are reading Marcel Proust's magnum opus or one of Ernest Hemingway's short stories. Some researchers have found that these discrepancies create enough so-called haptic dissonance to dissuade some people from using e-readers.

To amend this sensory incongruity, many designers have worked hard to make the e-reader or tablet experience as close to reading on paper as possible. E-ink resembles typical chemical ink, and the simple layout of the Kindle's screen looks remarkably like a page in a paper book. Likewise, Apple's iBooks app attempts to simulate somewhat realistic page turning. So far such gestures have been more aesthetic than pragmatic. E-books still prevent people from quickly scanning ahead on a whim or easily flipping to a previous chapter when a sentence brings to the surface a memory of something they read earlier.

Some digital innovators are not confining themselves to imitations of paper books. Instead they are evolving screen-based reading into something else entirely. Scrolling may not be the ideal way to navigate a text as long and dense as Herman Melville's *Moby-Dick,* but the *New York Times,* the *Washington Post,* ESPN, and other media outlets have created beautiful, highly visual articles that could not appear in print because they blend text with movies and embedded sound clips and depend entirely on scrolling to create a cinematic experience. Robin Sloan has pioneered the tap essay, which relies on physical interaction to set the pace and tone, unveiling new words, sentences, and images only when someone taps a phone or tablet's touch screen. And some writers are pairing up with computer programmers to produce ever more sophisticated interactive fiction and nonfiction in which one's choices determine what one reads, hears, and sees next.

When it comes to intensively reading long pieces of unembellished text, paper and ink may still have the advantage. But plain text is not the only way to read.

SARAH STEWART JOHNSON

O-Rings

FROM *Harvard Review*

BENEATH THE BLINDING WHITE SKY, where glaciers calve and crash into the Ross Sea and the land surface of Antarctica begins, there are two isolated huts, the Discovery and the Terra Nova. The Discovery hut was erected in 1902 at the dawn of the age of Antarctic exploration. British Royal Navy Captain Robert F. Scott picked up the prefabricated structure in Melbourne on his way south. No one gave much thought to the wide low-angle roof and broad windows, both designed to dissipate heat in the Australian outback. No one had expected to live in the hut or, in reality, to be stranded there. In desperation, the last inhabitants took to ripping down the ceiling. They burned the rafters, still lanced with nails, in exchange for a few hours of heat. The walls are smoke-stained and jagged.

The Terra Nova hut is 12 miles farther north. It was built in a little over a week nine years later and was used by Scott as the staging center for his second, doomed attempt at the pole. Its walls are double-planked, stuffed with lint and seaweed. It's attached to a set of stables designed for the expedition's nineteen Manchurian ponies, though none of them lasted very long on the ice. The insides of both of these huts remain perfectly intact—not because the structures have been made into museums but because nothing decays in the frigid cold and everything was left. If you go to Antarctica's research station in McMurdo Sound and you wrangle a key and a helicopter ride to Cape Evans, you'll find everything inside the Terra Nova hut just as it was when the members of Scott's party who didn't attempt to get to the pole went running for the ship to take them home. They left their possessions, their papers, even the dog whose skeleton you can still see, bound with a metal collar, on the floor of the doghouse.

These two huts lie near the research station where I spent the coldest summer of my life. I was twenty-six years old and in the middle of my graduate work as a planetary scientist. I went to Antarctica to probe for traces of life beneath the snow of its harsh, clean deserts. Regions of the continent were known for their similarities to Mars, which is why I had come, as an aspiring explorer of that distant planet. But I was also lured to Antarctica by something I'd once read by Edwin Mickleburgh, who wrote in *Beyond the Frozen Sea* that "its overwhelming beauty touches one so deeply that it is like a wound."

Each day in Antarctica I would rise and don the dozen layers of thermal underwear and goose down I had been issued. I was part of a program sponsored by the National Science Foundation to train young researchers. Some mornings we would snowmobile out over the windblown blue ice, a sledge of equipment fishtailing behind us. On other mornings we would head down to the helipad, hop into an A-Star or Bell 212, and zoom out over the booming pressure ridges. When the pilot reached the edge of the sea ice, one of us had to jump out, the helicopter still hovering, and bore a hole with a 4-inch bit. The ice had to be at least a meter thick; if not, it would buckle beneath us.

Everywhere I went, I lugged a gargantuan survival bag that weighed nearly as much as I did. Inside the waterproof red vinyl flaps were sleeping bags, a tent, and stakes, a WhisperLite stove, two quarts of white gas, a cookset, six freeze-dried meals, six candy bars, two bricks of Mainstay 3600 survival ration, tea bags, cocoa packets, toilet paper, candles, matches, a signal kit, and a standard-issue romance novel to read while waiting to be rescued. I practiced sawing building blocks out of the snow to construct ice walls, survival trenches, and snow caves. I studied the HF radio alphabet: *alpha, bravo, charlie, delta, echo.* I learned the geometry of crevasses and how never to step near where three cracks crossed, for the ice could give way like a trapdoor.

I also learned about the weather, which was constantly changing. A column of cold air could suddenly sink and roll over the terrain. Within seconds, a completely calm afternoon could be swallowed by katabatic winds howling off the East Antarctic ice sheet.

On overcast days, the white clouds could merge with the white snow; the light could become so diffuse that shadows disappeared, making it impossible to judge distances or distinguish the horizon.

Each of us had our own scientific interests: penguins, ciliates, the flapping valves of Antarctic scallops. For me it was the bacterial cells eking out a living in the bleakness of inland Antarctica. I analyzed samples from the Dry Valleys, just across the sound, where no rain has fallen for two million years. There iron oxide minerals, which also tint the surface of Mars, stain the blood-red tongue of a glacier that dips down to the ice-covered surface of a salt lake. I studied Bratina Island, which has all the indicators of land but isn't land at all, just a thick layer of dirt and rocks resting on a layer of ice floating, in turn, upon the sea. Slick mats of green, yellow, and orange cyanobacteria are suspended there like felt in the meltwater ponds, gashes of color against the barren terrain. I investigated all the microbial colonies I could find, trying to understand how pockets of life could survive in the hostile, Mars-like conditions.

Unexpectedly, though, it wasn't the continent's biology that most moved me, or its tumbling crevasses or poleward storms. Or even the remarkable extent to which my inner world flowed out into the landscape. It was those huts built by Scott and his two polar expedition parties. In contrast to the shimmering ice, the world inside them was dark and awful; there were reverberations from the walls, the abandoned tins and boxes, many of them still full. There were bottles of ketchup, tins of cabbage, a gramophone, test tubes, and glass vials with chemical powders. Ruined reindeer boots, man-hauling sled belts, stacked carcasses of seals, the echoes of death.

There's a small library at McMurdo, located between the laundry room and the weight room at the rear of Building 155. It has no windows and about thirty shelves of books. During the light-washed nights, when I couldn't sleep, I would sometimes find myself there, studying the faces of the men who once inhabited those huts. I would curl up on a piece of battered furniture and look through books and photographs archiving the early expeditions. One day I came across the diaries of Edward Wilson, the scientist of the crew. He last saw the Terra Nova hut in late October 1911, when he joined Scott on his final expedition. All five men in the party reached the South Pole only to discover that the Norwegian flag had been planted there a few weeks earlier, and all five men died on the return journey. Wilson collected thirty-five pounds of geologic fossils proving that Antarctica was once covered by ferns. The consummate scientist, he hauled those fossils to the very end.

On his way back from the pole, Wilson catalogued the ambient temperatures, which remain to this day among the coldest ever recorded on the South Polar Plateau. So cold that a glass of water thrown into the air would freeze before it hit the ground. A few days later the expedition unearthed a stored cache of supplies at Middle Barrier Depot only to discover that the canisters of fuel had evaporated. It was early March 1912, just a couple of weeks before their death. They needed fuel to melt drinking water and dry out their clothes. Without it, they slowly became encased in a mantle of frozen fabric. There was nothing to treat their frostbitten toes. No heat to draw them from their reindeer-skin sleeping bags in the morning. No warmth to help their shivering bodies to sleep at night. It was a major turning point in the expedition and, as it turns out, the evaporation of the fuel can be attributed to something very small. The O-rings, the flexing gaskets that acted to seal the fuel inside the canisters, turned brittle and cracked in the extreme cold.

As a space scientist, I know something about O-rings. In 1986, seventy-four years after Scott's party met its end, Caltech professor Richard Feynman sat before an investigative panel and dropped an O-ring into a glass of ice water to demonstrate how circles of rubber lose their pliability in freezing conditions. Afterward, he placed the O-ring down on the wooden podium, looked solemnly ahead, and said, "I believe this has some significance for our problem." And indeed it did; he was part of the committee of scientists reviewing the *Challenger* shuttle disaster.

I was six years old when the *Challenger* exploded, but I remember it well. Christa McAuliffe was going to be the first schoolteacher in space, and, like schoolchildren around the United States, I was peering up at a television watching the liftoff live. After a few moments, Mrs. Schrader walked to the front of the room, her face white, and clicked off the power. It's one of my earliest memories, and yet it didn't alter my desire to become an astronaut. Even now, even with two shuttles down and NASA's human space-flight program in disarray, I still think about soaring off in a rocket.

A hundred years have passed since Scott's expedition, and the frontier is now the void of outer space. Like many other young scientists, I have levied my striving upon this great unknown, but I sometimes worry that my convictions about exploration are inaccurate. What if the actuality of this enterprise is horribly different from my romantic ideas?

Inside the Terra Nova hut, I lingered by the bunk of Captain L. E. G. Oates. He was the second person to die on the way back from the South Pole, and his small space remains, to this day, cluttered with cavalry equipment. In the McMurdo library, I couldn't stop looking at the pictures of Oates taken by the expedition's photographer. In one of them, Oates is standing in the stables, now empty and stained with seal blood. The light falls gently on his right shoulder, and from beneath a thick wool hat, he looks intently at the camera.

Oates developed a savage case of frostbite on the return journey. With the winter cold and darkness descending, Scott described how Oates stepped from his tent into a minus-40-degree blizzard, simply remarking to the others, "I am just going outside and may be some time." Scott's description of the young captain disappearing into the whiteness, sock-footed and alone, offering up his life to save his comrades, echoed throughout Britain. After the world learned of the tragedy, the *Evening News* called for the story to be read to children across the nation.

Scott's account, however, is in marked contrast to Wilson's blunt and matter-of-fact telling of the grim narrative. Wilson writes on November 2, 1911, "Efforts were absurd . . . ," on December 18, "Our hunger is very excessive . . ." And when he writes about Oates, there is nothing to suggest that he died in a whirl of gallantry. In a letter Wilson wrote to Oates's mother describing the death of her son there's no mention of any heroic last acts. In fact, that story is only found in Scott's journal entry some days later. It appears, in fact, that Oates did nothing to mark the occasion. Captain Oates, Captain L. E. G. Oates, his cankered legs rotting, just stumbled outside, and no one tried to stop him. He wasn't a man preciously composed in his suffering, inviolable as he faced oblivion. In all likelihood, he was a desperate ghost, seething with anger toward Scott, cut loose by pain, and on the verge of insanity.

The account of what happened to the crew of the *Challenger* has similar discrepancies. The spacecraft was launched on a cloudless day from Cape Canaveral in January 1986. Liftoff time was 11:38 A.M. Shortly after, a brittle O-ring turned the shuttle into an inferno of flaming liquid oxygen. The *New York Times* headline the next day reported: "Challenger Shuttle Explodes Seventy-Four Seconds Into Launch: Seven Astronauts Killed Instantly."

But in late February, divers located the crew module, which had barreled into the sea floor. An NBC report indicates that as cables pulled the wreckage onto the deck of a ship, a blue protrusion slipped out, bobbed along, and then disappeared back into the sea. It was the waterlogged body of astronaut Gregory Jarvis, and it was another five weeks before divers relocated his corpse.

Slowly, the true story was pieced together from images, debris, and the recovered wreckage. It was determined that just over a minute after liftoff, the booster stack of the shuttle had exploded, some 48,000 feet above the earth. The forward fuselage, the small tip of the shuttle harboring the crew module, separated from the tanks. Chillingly, it did not explode. Propelled by its own momentum, it rose away from the fireball, carrying its seven passengers. It streaked across the sky along a ballistic trajectory, arcing in the tender thread of a parabola. What the astronauts would have noticed in the moments after the explosion was actually the quietness, the roar of the engines ceasing as the fuel tanks ripped apart and the fuselage broke away.

In the hush, the crew module continued to rise above the smoke, which blossomed like a white geranium. Twenty-five seconds after the O-ring gave way, the crew module crested at 65,000 feet, and then it began to fall. It fell for two minutes and forty-five seconds before impacting into the ocean. To the experts who sifted through the pixelated images, nothing suggested that it was erratically pitching or yawing. Among the most haunting pieces of wreckage discovered were four personal egress air packs designed to provide breathable air in the event of emergencies. Three of them had been activated, and the official report determined that they had been activated manually, not as part of the impact. For how many of those two minutes and forty-five seconds did the seven astronauts remain conscious? On the investigation committee, some scientists and fellow astronauts thought they were conscious for all of it, but we'll never really know.

That "we'll never really know" was the overriding message of the official report as it was released. Much of the hardware was mangled beyond recognition, and many pieces were not found. The section of the report dealing with what happened to the astronauts seems nebulous, buried in the middle, as the reader's attention is swept off to the fact of brittle rubber. The bobbing, waterlogged corpse of Gregory Jarvis is never discussed, nor are those two minutes and forty-five seconds and what they might have been like. They are seldom mentioned, out of what I take to be respect and reverence at the heart of our collective American narrative about what it was to lose the *Challenger.*

I have spent a lot of time trying not to think about those last couple of minutes. I'm not sure I want to know that Mike Smith saw the flames ripple over his window, or what he imagined in his last two minutes and forty-five seconds of thought. I think of all of the things I can think of in that amount of time. I don't want to dwell on that sudden, serene silence or let myself envision Judy Resnik frantically searching for her accessory oxygen tank. I don't want to know that Captain Oates died in a state of psychosis either. I try to reason with myself that the details of these particular narratives are in the end inconsequential. They died: that is the sad fact of it, let it be.

But somehow I can't stop wondering about them. Five people perished on their way back from the pole, and another seven on their way to orbit; they fell against the terrifying whiteness of the ice and the burning sky. All twelve of those lives ended in blackened flesh, horror, and numbness, and all because of the tiny, tragic fact of a brittle O-ring.

That summer when I was twenty-six, I was in Antarctica long enough to notice changes in the light. By the end of my field season, the midnight sun was beginning to dip down lower and lower toward the horizon, glowing with a faint, creamy incandescence. Before I boarded the cargo flight home, I was determined to hike out and see the view from Castle Rock, a distant volcanic outcrop jutting through a glacier.

Early one morning, I stopped by the firehouse at McMurdo Station to pick up a radio, file a foot plan, and check out with the responder on duty. I stuck close to the wind-frayed flags planted every 20 feet along the route. The day was staggeringly bright and, like everywhere in Antarctica, devoid of smell. There were snowfields in every direction, and I could hear my boots with each step, pressing into snow so cold and dry it squeaked like Styrofoam. When I reached the end of the ice, I began climbing, pulling myself up along the twisted preset ropes. My boots gripped the cold igneous rock below me, once lava in the vent of an ancient volcano. As I ascended onto the peak, I fell on my knees, exhausted and sweating inside my big red coat. I gazed across the Ross Ice Shelf toward the South Pole. Behind me towered Mount Erebus and Mount Terror. In front soared the Royal Society Range, and to the east, the open sea, riddled with tabular icebergs.

The whole of the sprawling base—the labs, the dorms, the helicopter pads—had disappeared into a bewildering sea of white. I thought of the photograph of Earth taken by *Voyager I* from the edge of the solar system. From 3.7 billion miles away, the whole of our world—everything human—was less than a pixel across. After a while, I spotted the trail leading back to McMurdo Station, just a faint strand tracking toward the horizon. The same year as the *Challenger* accident, two Americans fell into a fissure of ice 75 feet below. A search and rescue team tried frantically to pulley them to the surface, but they were wedged in too tightly to budge. The team heard them crying as the hours passed, then wailing and screaming until their voices finally stopped.

What is it that drives us to places like these—to the nothingness of the poles, the vast void of outer space? At the edge our world recedes, but we can't escape the brittle cold, the throbbing legs, the grating of mechanical parts, the absurdity of those O-rings. But perhaps this is why we strive. Perhaps, in the midst of such immensity, when we are faced with the irreducible fact of us, the firmities of reason and rationality give way. In the muscle of this great paradox, even a scientist is capable of believing in bigger things. The stronger the contradiction, the tauter the bow, the farther we can shoot.

Standing in the piercing air, I began to feel incredibly cold. My skin was damp, and snow had found its way into the crevices of my wrists. On top of Castle Rock, with my breath tumbling down the peak, I took one last look at the vast expanse that surrounded me. There was total stillness except for the faint whirr of a distant helicopter. The Terra Nova and Discovery huts were out there somewhere, frozen and timeless, holding steady against the winds. I took hold of the fraying rope at my feet and began my descent.

BARBARA J. KING

When Animals Mourn

FROM *Scientific American*

ON A RESEARCH VESSEL in the waters off Greece's Amvrakikos Gulf, Joan Gonzalvo watched a female bottlenose dolphin in obvious distress. Over and over again, the dolphin pushed a newborn calf, almost certainly her own, away from the observers' boat and against the current with her snout and pectoral fins. It was as if she wanted to nudge her baby into motion—but to no avail. The baby was dead. Floating under direct sunlight on a hot day, its body quickly began to decay; occasionally the mother removed pieces of dead skin and loose tissue from the corpse.

When the female dolphin continued to behave in this way into a second day, Gonzalvo and his colleagues on the boat grew concerned: in addition to fussing with the calf, she was not eating normally, behavior that could be risky for her health, given dolphins' high metabolism. Three other dolphins from the Amvrakikos population of about 150 approached the pair, but none disrupted the mother's behavior or followed suit.

As he watched the event unfold in 2007, Gonzalvo, a marine biologist at the Tethys Research Institute in Milan, Italy, decided he would not collect the infant's body to perform a necropsy, as he would usually have done for research purposes. "What prompted me not to interfere was respect," he told me earlier this year. "We were privileged to be able to witness such clear evidence of the mother-calf bond in bottlenose dolphins, a species that I have been studying for over a decade. I was more interested in observing that natural behavior than interrupting it by abruptly interfering and disturbing a mother who was already in obvious distress. I would define what I saw as mourning."

Was the dolphin mother truly grieving for her dead calf? A decade ago I would have said no. As a biological anthropologist who studies animal cognition and emotion, I would have recognized the poignancy of the mother's behavior but resisted interpreting it as mourning. Like most animal behaviorists, I was trained to describe such reactions in neutral terms such as "altered behavior in response to another's death." After all, the mother might have become agitated only because the strange, inert status of her calf puzzled her. Tradition dictates that it is softhearted and unscientific to project human emotions such as grief onto other animals.

Now, though, especially after two years' research for my book *How Animals Grieve*, I think Gonzalvo was correct in his judgment that the mother dolphin was mourning. In the past few years a critical mass of new observations of animal responses to death has bubbled to the surface, leading me to a startling conclusion: cetaceans, great apes, elephants, and a host of other species ranging from farm animals to domestic pets may, depending on circumstances and their own individual personalities, grieve when a relative or close friend dies. That such a broad range of species —including some quite distantly related to humans—lament the passing of loved ones hints that the roots of our own capacity for grief run very deep indeed.

Defining Grief

Since Charles Darwin's day, two centuries ago, scientists have debated hotly whether some animals display emotions beyond those associated with parental care or other aspects of survival and reproduction. Darwin thought that, given the evolutionary connection between humans and other animals, many emotions must be similar across species. He granted to monkeys, for instance, grief and jealousy, as well as pleasure and vexation. But the attribution of emotions such as these to animals fell increasingly out of mainstream scientific favor. By the early twentieth century the behaviorist paradigm held sway, with its insistence that only observable behavior of animals, not their interior lives, could be studied with rigor. Gradually the scientific embrace of animal emotion has revived, thanks originally in part to anecdotes from long-term field studies on large-brained mammals. From Tanzania, Jane Goodall recounted in heart-wrenching detail the young chimpanzee Flint's decline and death from grief only weeks after the death of his mother, Flo. From Kenya, Cynthia Moss reported that elephants attend to dying comrades and stroke the bones of deceased relatives. Field biologists and anthropologists began to ask questions about whether, and how, animals mourn.

To study and understand grief among animals, scientists need a definition that distinguishes it from other emotions. Whereas "animal response to death" embraces any behavior by an individual following the death of a companion animal, researchers may strongly suspect grief only when certain conditions are met. First, two (or more) animals choose to spend time together beyond survival-oriented behaviors such as foraging or mating. Second, when one animal dies, the survivor alters his or her normal behavioral routine—perhaps reducing the amount of time devoted to eating or sleeping, adopting a body posture or facial expression indicative of depression or agitation, or generally failing to thrive. For his part, Darwin conflated grief with sadness. But the two differ, primarily in intensity: the grieving animal is more acutely distressed, possibly for a more prolonged period.

This two-part definition is imperfect. For one thing, scientists lack a metric for evaluating exactly what counts as "more acutely distressed." Should the criteria for grief differ according to species, and might grief in other animals assume forms that are difficult for humans to recognize as mourning? The data are not yet available on these questions. Furthermore, mothers or other caretakers that constantly provide food or protection to infants that subsequently die cannot be said to have met the first criterion (going beyond survival-oriented behaviors), yet they remain among the strongest candidates for suffering survivor's grief.

Future studies of animal mourning will help refine this definition. For now, it furthers our critical assessment of responses made by animals when others around them die. For instance, baboon and chimpanzee mothers in wild African populations sometimes carry the corpse of the dead baby for days, weeks, or even months—a behavior that on the surface of things might look like grief. But they may not exhibit any significant outward indicator of agitation or distress. When the animals carry on with their routine be-

haviors, such as mating, their behavior does not meet the criteria
for mourning.

A Menagerie of Mourners

A wide range of species do exhibit behaviors that fit the two-part
definition of grief, however, elephants among them. A particularly compelling example of elephant mourning comes from Iain
Douglas-Hamilton of Save the Elephants and his team at Kenya's
Samburu National Reserve, who in 2003 tracked elephants' responses to the dying matriarch called Eleanor. When Eleanor col-
lapsed, a matriarch named Grace from another elephant family
immediately came to her aid, using her tusks to support Eleanor
back onto her feet. When Eleanor fell again, Grace stayed with
her, pushing on her body, for at least an hour, even though her
own family moved on. Then Eleanor died. During the course of
the week that followed, females from five elephant families, including Eleanor's own, showed keen interest in the body. Some
individuals appeared upset, pulling at and nudging the body with
trunk and feet or rocking back and forth while standing over it.
Based on the females' reactions (at no point during this period
did a bull elephant visit the carcass), Douglas-Hamilton concluded
that elephants show a so-called generalized response to dying and
death—grieving not only for the loss of close kin but for individuals in other families.

Wild cetaceans also seem to exhibit a generalized grief response. In the Canary Islands in 2001, Fabian Ritter of Mammal
Encounters Education Research observed a rough-toothed dolphin mother pushing and retrieving her dead calf's body in much
the same way that the Amvrakikos dolphin mother had done with
her baby's corpse. She was not alone: two adult escorts swam syn-
chronously with her at certain periods, and at other times a group
of at least fifteen dolphins altered their pace of travel to include
the mother and dead baby. The mother's persistence was remarkable, and when on the fifth day it began to wane, the escorts joined
in and supported the infant on their own backs.

Giraffes, too, appear to grieve. In 2010 at the Soysambu Conser-
vancy in Kenya, a female Rothschild's giraffe gave birth to a baby

with a deformed foot. The baby walked less and remained more stationary than most calves. During the youngster's four weeks of life, wildlife biologist Zoe Muller of the Rothschild's Giraffe Project, based in Kenya, never saw the mother more than 20 meters away. Although individuals in a giraffe herd often synchronize their activities, foraging together, for example, the mother deviated from this pattern, preferring to stay close to the baby. Like the dolphin mother in the Amvrakikos Gulf, she may have risked her own health in doing so—though in this case for a living offspring.

One day Muller discovered the herd engaged in highly atypical behavior. Seventeen females, including the calf's mother, were vigilant and restless as they stared into a patch of bush. The calf had died in that spot about an hour before. All seventeen females showed keen interest in the body that morning, approaching and then retreating from it. By the afternoon twenty-three females and four juveniles were involved, and some nudged the carcass with their muzzles. That evening fifteen adult females clustered closely around the body—more closely than they had been during the day.

Throughout the following day numerous adult giraffes attended the infant's body. Some adult males approached for the first time, although they showed no interest in the carcass, instead focusing on foraging or inspecting the reproductive status of the females. On day three Muller spotted the mother giraffe alone under a tree about 50 meters from where the calf had died. The body itself, however, was no longer in its resting spot. Following a search, Muller located it, half devoured, in the spot under the tree where the mother had been earlier. By the next day the body was gone, taken by hyenas.

Giraffes are highly social animals. After caching a newborn out of sight for about the first four weeks of life, the mothers sometimes engage in a crèche system in which one looks after the infants while the others forage. Muller does not use the words "grief" or "mourning" in describing the incident she witnessed. Yet this case is especially instructive. Not only the mother's behavior but also that of many of the females in her herd changed significantly in the wake of the infant's death. Although it is impossible to rule out an alternative explanation, the fact that the females had mounted a protective response against predators taking the baby makes it overwhelmingly likely that grief was involved at some level.

Detailed observations of wild populations of animals, such as the ones Muller reported, are still relatively rare, for several reasons. Scientists may not be at the right place at the right time to observe post-death responses by survivors. And even when they are present, no remarkable grief behaviors may ensue. Especially at this early stage of research into animal grief, observations from sanctuaries, zoos, and even our own homes may supply needed clues.

I cannot imagine describing the behavior of Willa the Siamese cat without invoking the word "grief." For fourteen years Willa lived with her sister, Carson, at the home of Karen and Ron Flowe in Virginia. The feline siblings groomed each other, lazed together in favorite parts of the house, and slept with their bodies entwined. If Carson was taken from the house to visit the vet, Willa acted mildly agitated until she reunited with her sister. In 2011 Carson's chronic medical issues worsened, and the Flowes took her again to the vet, where she died in her sleep. At first Willa acted as she did when her sister was away for a brief period. Within two or three days, though, she began to utter an unearthly sound, a sort of wail, and to search the spots she and Carson had favored together. Even when this startling behavior faded, Willa remained lethargic for months.

Of all the instances of animal grief I have compiled, the most surprising came from a sanctuary setting. In 2006 three mulard ducks arrived at Farm Sanctuary in Watkins Glen, New York. They suffered from hepatic lipidosis, a liver disease caused by force-feeding of the birds at a foie gras farm. Two of the rescued ducks, Kohl and Harper, were in bad shape physically and emotionally. Very afraid of people, Kohl had deformed legs, and Harper was blind in one eye. The two forged a fine supportive friendship for four years. Ducks are social birds, but even so, the intensity of their bond was unusual. When Kohl's leg pain increased and he could no longer walk, he was euthanized. Harper was allowed to observe the procedure and to approach his friend's body afterward. After pushing on the body, Harper lay down and put his head and neck over Kohl's neck. There he stayed for some hours. In effect, Harper never recovered from his loss. Day after day, he snubbed other potential duck friends, preferring to sit near a small pond where he had often gone with Kohl. Two months later Harper died as well.

The Sorrow Continuum

It is logical to think that long-lived species whose members partner most closely with others in tight-knit pairs, family groups, or communities may more readily mourn the deaths of loved ones than other species do. But researchers do not yet know enough about animal grief to make such a claim. We need to test this hypothesis by systematically comparing responses to death in a variety of animal social systems, from gregarious ones to those in which animals come together only seasonally for food or mating.

Still, species-level differences in grieving will not be the whole story, because variation in the immediate social contexts and personalities of individual survivors will complicate matters. For instance, whereas the practice of allowing a survivor to view the body, as Harper did with Kohl, sometimes seems to prevent or reduce a period of distressed searching and vocalizing by the surviving animal, at other times it seems not to help at all—attesting to the degree of individual variation in death responses within species. Likewise, evidence for grief in wild monkeys that live in cohesive social units is surprisingly limited so far, whereas in more solitary species such as domestic cats, bonds may develop between two or more kin or friends such that grief responses rival those of much more social animals. I would predict that field observations will show that some monkeys across varied social systems visibly mourn as much as some domestic cats. Indeed, in *How Animals Grieve*, I recount examples from cats, dogs, rabbits, horses, and birds, as well as the other animals discussed here. In each species I find a grief continuum, with some individuals seeming indifferent to a companion's death and other individuals appearing distraught over such a loss.

Cognitive differences also play a role in animal grief. Just as there are different levels of empathy expressed by different species and even across individuals within a species, there must be varying levels of comprehension when animals grieve. Do some animals grasp death's finality or even have a mental concept of death? We simply don't know. No evidence suggests that any nonhuman animal anticipates death in the way we humans do, a capacity that underlies so much of our compelling literature, music, art, and theater—and that costs our species a great deal in terms of emotional suffering.

Indeed, the capacity to mourn may become quite costly for any animal in both physical and emotional terms, especially in the wild, where alert high-energy behavior is needed for foraging, predator avoidance, and mating. Why then did grief evolve in the first place? Perhaps the social withdrawal that often accompanies an animal's grief, if not taken too far, allows time for rest and thus an emotional recovery that in turn leads to greater success in forging a new close bond. Or, as John Archer writes in *The Nature of Grief*, it may be that "the costs involved in grief can be viewed as a trade-off with the overall benefits conferred by separation responses" seen when two individuals are keenly attached but forced apart from each other. Under such circumstances, the missing partners may search for each other and thereby reunite and live to see another day. What is adaptive, then, may not be grief itself but instead the strong positive emotions experienced before grief comes into the picture, shared between two or more living animals whose level of cooperation in nurturing or resource-acquisition tasks is enhanced by these feelings.

The Price of Love

From this perspective, we may link grief with love, full stop. That is to say, grief results from love lost. Exploring emotions in a variety of species, ecologist and animal behaviorist Marc Bekoff of the University of Colorado at Boulder embraces the idea that many animals feel "love" as well as "grief," even as he acknowledges that those concepts are hard to define precisely. We humans, he notes, do not fully understand love, but we do not deny its existence—or its power to shape our emotional responses.

In his book *Animals Matter*, Bekoff tells the story of a coyote called Mom whom he observed for several years during behavioral studies in Wyoming's Grand Teton National Park. At one point Mom began to make short journeys on her own away from her pack. Her offspring would rejoice when she returned: they licked Mom and rolled over exuberantly at her feet. Then Mom left for good. Some of the coyotes in her pack paced; others searched for her, setting off in the direction Mom had taken. "For more than a week some spark seemed to be gone," Bekoff writes. "Her family missed her." Discussing animal emotion with me earlier this year, Bekoff attributed the family's response to its love for Mom. Generally, the potential for love is strong in species such as coyotes, wolves, and many birds, including geese, he said, because male and female partners defend territories, feed and raise their young together, and miss each other when they are apart.

Love in the animal world often entwines with grief in an acute mutuality. Perhaps even more than the degree of social cohesion within a species, it is love between individuals that predicts when grief will be expressed. Can there be any real doubt that Willa, a representative of a species (the domestic cat) not known for its social nature, loved her sister, Carson, or that as the sole surviving sister, she suffered grief in the wake of her loss?

In our own species, grief increasingly became expressed through rituals rich in symbolism. By around 100,000 years ago, our *Homo sapiens* ancestors decorated dead bodies in red ocher, a behavior interpreted by archaeologists to be a kind of symbolic (rather than functional) ornamentation. At a site in Russia called Sunghir, two children younger than thirteen years, a boy and a girl, were buried 24,000 years ago, together with elaborate grave goods ranging from mammoth tusks to animals carved from ivory. Most astonishing were the thousands of ivory beads found in the pair's grave, probably sewn onto the clothing (long since disintegrated) in which the children were buried. A good portion of this ancient human community at Sunghir must have come together in preparing this funeral ritual—each bead alone took an hour or more to manufacture. Although it is risky to project modern emotions onto past populations, the examples of animal grief reviewed here strengthen an emotion-based interpretation of the archaeological evidence: our ancestors of many thousands of years ago mourned their lost children.

In our modern world, grief is no longer inevitably confined to kin, close social partners, or immediate members of one's own community. Public commemoratives at the Peace Memorial Park in Hiroshima; the genocide memorial center in Kigali, Rwanda; the Foundation Memorial to the Murdered Jews of Europe in Berlin; or the site of the Twin Towers in Manhattan or Sandy Hook

Elementary School in Newtown, Connecticut, all convey visibly the power of agonized global mourning. Our uniquely human capacity for sorrow at the deaths of those who are strangers to us is built on an evolutionary substrate. Our own ways of mourning may be unique, but the human capacity to grieve deeply is something we share with other animals.

BARBARA KINGSOLVER

Where It Begins

FROM *Orion*

IT ALL STARTS with the weather. Comes a day when summer finally gives in to the faintest freshet of chill and a slim new light, and just like that, you're gone. Wild in love with the autumn proviso. You can see that the standing trees are all busy lighting themselves up ember-orange around the hemline, starting their ritual drama of slow self-immolation—oh, well, you see it all. The honkling chain gang of boastful geese overhead that are fleeing warmward-ho, chuckling over their big escape. But not you. One more time, here for the duration, you will stick it out. Through the famously appley wood-smoked season that opens all hearts' doors into kitchen industry and soup on the stove, the signs wink at you from everywhere: sticks of kindling in the fire, long white brush-strokes of snow on the branches, this is the whole world calling you to take up your paired swords against the brace of the oncoming freeze. The two-plied strands of your chromosomes have been spun by all thin-skinned creatures for all of time, and now they offer you no more bottomless thrill than the point-nosed plow of preparedness. It begins on the morning you see your children's bare feet swinging under the table while they eat their cereal cold, and you shudder from stem to stern like a dog hauling up from the lake, but you can't throw off the clammy pall of those little pink-palmy feet. You will swaddle your children in wool, in spite of themselves.

It starts with a craving to fill the long evening downslant. There will be whole wide days of watching winter drag her skirts across the mud-yard from east to west, going nowhere. You will want to nail down all these wadded handfuls of time, stick-pin them to the blocking board, frame them on a twenty-four-stitch gauge. Ten to the inch, ten rows to the hour, straggling trellises of days held fast in the acreage of a shawl. Time by this means will be domesticated and cannot run away. You pick up sticks because time is just asking for it, already lost before it arrives, scattering trails of leavings. The frightful movie your family has chosen for Friday night, just for instance. They insist it will be watched, and so with just the one lamp turned on at the end of the sofa you can be there too, keeping your hands busy and your eyeshades half drawn. Yes, people will be murdered, cars will be wrecked, and you will come through in one piece, plus a pair of mittens. It's all the same wherever you go—the river is rife with doldrums and eddies, the waiting room, the plane, the train, the learned lecture, the meeting. Oh, sweet mother of Christ, the meeting. The PTA the town council the school board the bored-board, the interminably haggled items of the agenda. Your feet want to run for their lives, but your fingers know to dig in the bag and unsheathe their handy stays against impatience, the smooth paired oars, the sturdy lifeboat of yarn. This giant unwieldy meeting may bottom-drag and list on its keel, stranded in the Sargasso Sea of Agenda, but you alone will sail away on your thrifty raft of unwasted time. You alone are swaddling the world in wool.

Strangely, it also begins with the opposite: a hankering to lose time and all sense of purpose. To banish all possibilities, the winter and the summer, the bare feet under the table, the shattered day undone, and dregs of old regard and bitter unsettled tea leaves, and the words forever jostling ahead of each other in line, queuing up to be written. Especially those. Words that drub, drub, drub at the skull's concave inner wall. Words that are birds in a linear flock, pelting themselves in ruined fury all night long against the windowpane. Nothing can stop the words so well as the mute alphabet of knit and purl. The curl of your cupped hand scoops up long drinks of calm. The rhythm you find is from down inside, rocking cradle, heartbeat, ocean. Waves on a rockless shore.

Sometimes it starts terribly. With the injury or the accident or the wrecked life flung down like an armload of broken chair legs on your doorstep. Here lies the recuperation, whose miles you can't even see across, let alone traverse. Devil chasm of woe uncrossable by any known bridge. And in comes the friend bearing needles of blond bamboo—twin shafts of light!—and ombré skeins in graded shades that march through the stages of grief, burnt umber to ocher to gold to dandelion. She is not in a listening mood, the friend. Today she commands you to make something of all this. And to your broken heart's surprise, you do.

It begins with the circle of friends. There is always something beyond your beyond, the aged parents and teenager who crack up the family cars on the selfsame day, the bone-picked divorce, the winter of chemo, the gorgeous mistake, the long unraveling misery that needs company, reading glasses and glasses of wine and all the chairs pulled into the living room. Project bags bulge like sacks of oranges, ripe for beginning. Cast on, knit two together girlfriendwise. Rip it, pick up the pieces where you can, along the headless yoke or scandalously loose button placket, pick up and knit. Always, you will have to keep two projects going: first, the no-brainer stockinette that can run on cruise control when the talk is delicious. And the other one, the brainer, a maddening intarsia or fussy Fair Isle you'll save for the day when the chat gets less interesting, though really it never does. Knitting only makes the talk go softer, as long as it needs to be, fondly ribbed and yarned-over, loosely structured or not at all, with embellishment on every edge. Laughter makes dropped stitches.

It begins with a pattern. The arresting helical twist of a double cable, a gusset, a hexagon, a spiral, a fractal, an openwork ladder, an Aran braid, a chevron and leaf, the eyes of the lynx, the traveling vines. The mimsy camisole you arguably could live without, the munificent cardigan you need. A mitten lost in childhood, returned to you in a dream. A pattern in a magazine, devised of course to tantalize. More embarrassing yet, the pattern hallooing from your neighbor's sweater while you're only trying for small talk, distracting you until finally you have to stop, apologize, and ask permission to stare and memorize the lay of her sweater's land. And once it all starts, there's no stopping. The frame of your four double-points is a sturdy raised bed from which you cultivate the lively apical stem of sock-sleeve-stocking-cap. It's all in the growing. From the seed of pattern, the cotyledons of cast-on, everything rises: xylem and phloem of knit-purl ribs, a trunk of body and branches of sleeves, the skirt that bells downward daffodilwise. You with your needles are god of this wild botany. It begins the first time you take the familiar map in hand, scowling it over with all best intentions, then throw it over your shoulder and head out to uncharted waters where there be monsters. Only there will you ever discover the promised land of garments heretofore undevised. Gloves for the extra long of hand, or short, or the firecracker nephew with one digit missing in action. Sweaters for the short-waisted, the broad-shouldered, the precise petite. Soon they are lining up, friends and family all covetous of the bespoke, because your best beloveds are human after all, and not off-the-rack. You can envelop each of them in the bliss of a perfect fit.

And a perfect color. It starts there too. Every eye has hungers all its own. The particular green-silver of leaves overturned by the oncoming storm. An alkaline desert's russet bronze, a mustard of Appalachian spring, some bright spectral intangible you find you long to possess. Colors are fertilized in vitro with the careful spoon and the potent powder weighed to the iota, and born by baptism in the big dye kettle hauled onto the stove. Flaccid beige hanks backstroke listlessly in the boiling ink, waiting to be born again, until some perfect storm of chemical zeal moves them suddenly to awaken and drink down all the dye molecules in a trice. Like a miracle, the dark liquid goes clear as water before your very eyes. Afterward the damp yarn sings its good news from dripping loops in the laundry room, waiting to meet the pattern the wish the cool weather the living room the days-long patient fortune.

It starts with a texture. There are nowhere near enough words for this, but fingers can sing whole arpeggios at a touch. Textures have their family trees: cloud and thistledown are cousin to catpelt and earlobe and infantscalp. Petal is also a texture, and limepeel and nickelback and nettle and five-o'clock-shadow and sandstone and ash and soap and slither. Drape is the child of loft and crimp; wool is a stalwart crone who remembers everything, while emptyhead white-haired cotton forgets. And in spite of their various natures, all these strings can be lured to sit down together and play a fiber concerto whole in the cloth. The virgin fleece of an April lamb can be blended and spun with the fleece of a fat blue hare or a twist of flax, anything, you name it, silkworm floss or twiny bamboo. Creatures never known to converse in nature can be introduced and then married right on the spot. The spindle is your altar, you are the matchmaker, steady on the treadle, fingers plying the helices of a beast and its unlikely kin, animal and vegetable, devising your new and surprisingly peaceable kingdoms. Fingers can coax and read and speak, they have their own secret libraries and illicit affairs and conventions. Twined into the wool of a hearty ewe on shearing day, hands can read the history of her winter: how many snows, how barren or sweet her mangers. For best results, stand in the pasture and throw your arms around her.

Because, really, it does start there, in the barn on shearing day. The circle of friends again, assembled for shearing and skirting. One whole fleece, shorn all of a piece, is flung out on a table like a picnic blanket, surrounded by women. All hands point toward the center like an excessive, introverted clock, the better for combing the white fleece with all those fingers; combing the black, fingers can see in the dark to pull out twigs and manure tags and cockleburs. White fleeces shaken free of second cuts, rolled and bundled and stacked, ready for spinning, look for all the world like loaves of bread on a bakery shelf, or sheaves of grain or any other money in the bank. The universal currency of a planet where people grow cold. On shearing day all ledgers will be balanced, the sheep lined up in the gates are woolly by morning and naked by night, as the barrows fill and the spindles make ready and warmth is bankrolled in futures. Six women can skirt a fleece in ten minutes, just enough time to run and collect the next one, so long as the shearer is handy. It starts early, this day, and goes long.

It starts in the barn on other days too, every morning of the year, in fact. The sheep are both eager and wary at the sight of you, the bringer of hay, the reaper of wool, as you enter the barn for the daily accounts. You switch on the overhead bulb and inhale the florid scents of sweet feed and hay and mineral urine, and there they stand all eyeing you with horizontal pupils, reliably here for every occasion, the blizzard nights and early spring mornings of lambing. You hurry out at dawn to find dumbfounded mothers of twins licking their wispy trembling slips of children, exhorting them to look alive in the guttural chortle that only comes into the throat of a ewe when she's just given birth. The sloe-eyed flock mistrusts you fundamentally, but still they will all come running when you shake the exquisite bucket of grain, the money that talks to yearlings and chary wethers alike, and loudest of all to the ravenous barrel-round pregnant ewes. They gallop home with their udders tolling like church bells. In all weather you take their measure and send them out again to the pasture. And oh, how willingly they return to their posts, with their gentle gear-grinding jaws and slowly thickening wool under winter's advance, beginning your sweater for you at the true starting gate.

Everything starts, of course, with the sheep and the grass. Beneath her greening scalp the earth frets and dreams, and knits herself wordless. Between her breasts, on all hillsides too steep for the plow, the sheep place little sharp feet on invisible paths and lead their curly-haired sons and daughters out onto the tart green blades of eternal breakfast. It starts on tumbled-up lambspring mornings when you slide open the heavy barn door and expel the pronking gambol of newborn wildhooray into daylight. And in summer haze when they scramble up onto boulders and scan the horizon with eyes made to fit it just-so, horizontal eyes, flattened to that shape by the legions of distant skulking predators avoided for all of time. And in the gloaming, when the ewes high up on the pasture suddenly raise their heads at the sight of you, conceding to come down as a throng in their rocking-horse gait, surrendering under dog-press to the barn-tendered mercy of nightfall. It starts where everything starts, with the weather. The muffleblind snows, the dingle springs, the singular pursuit of cud, the fibrous alchemy of the herd spinning grass into wool. This is all your business. Hands plunged into a froth of yarn are as helpless as hands thrust into a lover's hair, for they are divining the grass-pelt life of everything: the world. The sunshine, heavenly photosynthetic host, sweet leaves of grass all singing the fingers electric that tingle to brace the coming winter, charged by the plied double helices of all creatures that have prepared and justly survived on the firmament of patience and swaddled children. It's all of a piece. All one thing.

MAGGIE KOERTH-BAKER

Danger! This Mission to Mars Could Bore You to Death!

FROM *The New York Times Magazine*

RIGHT NOW, SIX PEOPLE are living in a nearly windowless, white geodesic dome on the slopes of Hawaii's Mauna Loa volcano. They sleep in tiny rooms, use no more than eight minutes of shower time a week, and subsist on a diet of freeze-dried, canned, or preserved food. When they go outside, they exit through a mock air lock, clad head to toe in simulated space suits. The dome's occupants are playing a serious version of the game of pretend—what if we lived on Mars?

Research at the Hawaii Space Exploration Analog and Simulation (HI-SEAS) project, funded in part by NASA, is a continuation of a long history of attempts to understand what will happen to people who travel through outer space for long periods of time. It's more than a technical problem. Besides multistage rockets to propel a spacecraft out of Earth's atmosphere, years of planning and precise calculations, and massive amounts of fuel, traveling the tens of millions of miles to Mars will take a tremendous amount of time. With current technology, the journey takes more than eight months each way.

Which means that astronauts will get bored. In fact, a number of scientists say that—of all things—boredom is one of the biggest threats to a manned Mars mission, despite the thrill inherent in visiting another planet. And so, attention is being paid to the effects of boredom at HI-SEAS and on the International Space Station. But because of the causes of chronic boredom, scientists say, research facilities in Antarctica might actually provide a better simulation of the stress of a journey to Mars.

Most living things constantly seek out sensory stimulation—new smells, tastes, sights, sounds, or experiences. Even single-celled amoebas will move to investigate new sources of light or heat, says Sheryl Bishop, who studies human performance in extreme environments at the University of Texas Medical Branch. Animals deprived of naturalistic environments and the mental stimulation that comes with them can fall into repetitive, harmful patterns of behavior. Anybody of a certain age will remember zoos full of maniacally pacing tigers, bears gnawing on their metal cages, and birds that groomed themselves bald—all a result, we now know, of their rather unstimulating lifestyles.

Human boredom isn't quite as well understood, says James Danckert, a professor of cognitive neuroscience at the University of Waterloo. He's currently working on what he says may be the first study of how our brain activity changes when we're bored. Danckert is hoping to find out whether boredom is connected to a phenomenon called the "default network"—a background hum of brain activity that seems to remain on even when you aren't directly focused on something. There's a lot of observable activity in the brains of people who are staring at a blank screen—way more than anybody expected, Danckert says. The default network maps closely to the brain-activity patterns scientists see when someone's mind is wandering. It suggests that what we call a restless mind is just that—a mind desperate for something to amuse it, searching frantically for stimulus.

Boredom, it turns out, is a form of stress. Psychologically, it's the mirror image of having too much work to do, says Jason Kring, president of the Society of Human Performance in Extreme Environments, an organization that studies how people live and work in space, underwater, on mountaintops and other high-risk places. If your brain does not receive sufficient stimulus, it might find something else to do—it daydreams, it wanders, it thinks about itself. If this goes on too long, it can affect your mind's normal functioning. Chronic boredom correlates with depression and attention deficits.

Astronaut candidates go through two years of training before they're even approved to fly. And before they are chosen to be candidates, they have to compete against thousands of other applicants. The 2013 class, for instance, had more than six thousand applicants, and only eight were chosen. Astronauts are rigorously tested for psychological as well as physical fitness. But no mission in NASA's history has raised the specter of chronic boredom to the degree that a Mars mission does, because none has involved such a long journey through nothingness.

What if, millions of miles from home, a chronically bored astronaut forgets a certain safety procedure? What if he gets befuddled while reading an oxygen gauge? More important, Danckert and Kring say, bored people are also prone to taking risks, subconsciously seeking out stimulation when their environment bores them.

The cognitive and social psychologist Peter Suedfeld says that people will sometimes do reckless, stupid things when they suffer from chronic boredom. In Antarctica, where winter can cut scientists and crew off from the rest of the world for as long as nine months, the isolation can lead to strange behavior. Suedfeld told me he has heard about Antarctic researchers venturing outside in 40-below weather without proper clothing and without telling anyone else they were going out.

The diaries of early polar explorers are full of tales of extreme boredom, depression, and desperate attempts at entertainment reminiscent of prisoners' stories from solitary confinement. An important lesson that Antarctica can impart on a Mars expedition is this: even scientists on important missions can get excruciatingly bored.

One effective way astronauts combat boredom is by staying busy with work. That's a strategy at HI-SEAS, where the crew member Kate Greene told me that her schedule is packed—every hour planned and accounted for, from the time she wakes up to the time she goes to bed at night. Life on the International Space Station is similar. (In fact, historically, NASA's problem has been overworking people: in 1973, the exhausted crew of Skylab 4 actually staged a relaxation rebellion and took an unscheduled day off.) But Antarctica is different from HI-SEAS or the International Space Station. Communications are limited. There's nobody outside the base directing your day. Spectacular views vanish in a haze of white. It's just you, the people you came in with, no way out, and little to break up the monotony.

And so some researchers there have learned to actively fend off boredom by creating what you might call a unique office culture. They celebrate a ridiculous number of holidays, both traditional and invented. You need something to look forward to, Suedfeld says, and planning the events helps change the routine. Even Ernest Shackleton's Antarctic crew found ways to put on skits and concerts. On one expedition, Shackleton brought a small printing press. At McMurdo Station, the 1983 winter crew created costumes, learned lines, and acted out scenes from the movie *Escape From New York*. It's possible that we may someday watch recordings of Mars-bound astronauts acting out other John Carpenter films. (It's not so far-fetched. Chris Hadfield, a Canadian astronaut, made a tribute to David Bowie's *Space Oddity* that racked up more than 16 million views on YouTube.)

It might sound absurd, but many scientists say strategies like this are necessary because without proper mental stimulus, we risk making a physically and technologically challenging endeavor into a psychologically grueling one. It would be catastrophic if humanity's greatest voyage were brought low by the mind's tendency to wander when left to its own devices.

ELIZABETH KOLBERT

The Lost World

FROM *The New Yorker*

1

On April 4, 1796—or, according to the French Revolutionary calendar in use at the time, 15 Germinal, Year IV—Jean-Léopold-Nicholas-Frédéric Cuvier, known, after a brother who had died, simply as Georges, delivered his first public lecture at the National Institute of Science and Arts in Paris. Cuvier, who was twenty-six, had arrived in the city a year earlier, shortly after the end of the Reign of Terror. He had wide-set gray eyes, a prominent nose, and a temperament that a friend compared to the exterior of the earth —generally cool, but capable of violent tremors and eruptions. Cuvier had grown up in a small town on the Swiss border and had almost no connections in the capital. Nevertheless, he had managed to secure a prestigious research position there, thanks to the passing of the ancien régime, on the one hand, and his own sublime self-regard, on the other. An older colleague later described him as popping up in the city "like a mushroom."

For his inaugural lecture, Cuvier decided to speak about elephants. Although he left behind no record to explain his choice, it's likely that it had to do with loot. France was in the midst of the military campaigns that would lead to the Napoleonic Wars and had recently occupied Belgium and the Netherlands. Booty, in the form of art, jewels, seeds, machinery, and minerals, was streaming into Paris. As the historian of science Martin J. S. Rudwick relates, in *Bursting the Limits of Time* (2005), a hundred and fifty crates' worth was delivered to the city's National Museum of Natural History. Included among the rocks and dried plants were two elephant skulls, one from Ceylon—now Sri Lanka—and the other from the Cape of Good Hope, in present-day South Africa.

By this point, Europe was well acquainted with elephants; occasionally one of the animals had been brought to the Continent as a royal gift, or to travel with a fair. (One touring elephant, known as Hansken, was immortalized by Rembrandt.) Europeans knew that there were elephants in Africa, which were considered to be dangerous, and elephants in Asia, which were said to be more docile. Still, elephants were regarded as elephants, much as dogs were dogs, some gentle and others ferocious. Cuvier, in his first few months in Paris, had examined with care the plundered skulls and had reached his own conclusion. Asian and African elephants, he told his audience, represented two distinct species.

"It is clear that the elephant from Ceylon differs more from that of Africa than the horse from the ass or the goat from the sheep," he declared. Among the animals' many distinguishing characteristics were their teeth. The elephant from Ceylon had molars with wavy ridges on the surface, "like festooned ribbons," while the elephant from the Cape of Good Hope had teeth with ridges arranged in the shape of diamonds. Looking at live animals would not have revealed this difference, as who would have the temerity to peer at an elephant's molars? "It is to anatomy alone that zoology owes this interesting discovery," Cuvier said.

Having successfully sliced the elephant in two, Cuvier continued with his dissection. Over the decades, the museum had acquired a variety of old bones that appeared elephantine. These included a three-and-a-half-foot-long femur, a tusk the size of a jousting lance, and several teeth that weighed more than five pounds each. Some of the bones came from Siberia, others from North America. Cuvier had studied these old bones as well. His conclusions, once again, were unequivocal. The bones were the fragmentary remains of two new species, which differed from both African and Asian elephants "as much as, or more than, the dog differs from the jackal." Moreover—and here one imagines a hush falling over his audience—both creatures had vanished from the face of the earth. Cuvier referred to the first lost species as a mammoth, and the second as an "Ohio animal." A decade later, he would invent a new name for the beast from Ohio; he would call it a mastodon.

"What has become of these two enormous animals of which one no longer finds any living traces?" Cuvier asked his audience. The question was more than rhetorical. Just a few months earlier, Cuvier had received sketches of a skeleton that had been discovered in Argentina. The skeleton was twelve feet long and six feet high; the sketches showed it to have sharp claws, flattish feet, and a short muzzle. On the basis of the sketches, Cuvier had identified its owner—correctly—as an oversized sloth. He named it *Megatherium,* meaning "great beast." Though he had never been to Argentina or, for that matter, anywhere farther than Stuttgart, Cuvier was convinced that the *Megatherium* was no longer to be found lumbering through the jungles of South America. It, too, had disappeared. Like the mammoth's and the mastodon's, its bones hinted at events both strange and terrible. They "seem to me," Cuvier said, "to prove the existence of a world previous to ours, destroyed by some kind of catastrophe."

Extinction may be the first scientific idea that children today have to grapple with. We give one-year-olds dinosaurs to play with, and two-year-olds understand, in a vague sort of way, at least, that these small plastic creatures represent very large animals that once existed in the flesh. If they're quick learners, kids still in diapers can explain that there were once many kinds of dinosaurs and that they lived long ago. (My own sons, as toddlers, used to spend hours over a set of dinosaurs that could be arranged on a plastic mat depicting a forest from the Cretaceous. The scene featured a lava-spewing volcano, and when you pressed the mat in the right spot it emitted a delightfully terrifying roar.) All of which is to say that extinction strikes us as an extremely obvious idea. It isn't.

Aristotle wrote a ten-book *History of Animals* without considering the possibility that animals actually had a history. Pliny's *Natural History* includes descriptions of animals that are real and animals that are fabulous, but no descriptions of animals that are extinct. The idea did not crop up during the Middle Ages or during the Renaissance, when the word "fossil" was used to refer to anything dug up from the ground (hence the term "fossil fuel"). During the Enlightenment, the prevailing view was that every species was a link in a great, unbreakable "chain of being." As Alexander Pope put it in his *Essay on Man:*

> All are but parts of one stupendous whole,
> Whose body nature is, and God the soul.

When Carolus Linnaeus introduced his system of binomial nomenclature, he made no distinction between the living and the dead, because, in his view, none was required. The tenth edition of his *Systema Naturae,* published in 1758, lists sixty-three species of scarab beetle, thirty-five species of cone snail, and fifteen species of flat fish. And yet in the *Systema Naturae* there is really only one kind of animal—those that exist.

This view persisted despite a growing body of evidence to the contrary. Cabinets of curiosity in London, Paris, and Berlin were filled with traces of strange marine creatures that no one had ever seen—the remains of what would now be identified as trilobites, belemnites, and ammonites. Some of the last were so large that their fossilized shells approached the size of wagon wheels. But the seas were vast and mostly unexplored, and so it was assumed that the creatures must be out there somewhere.

With his lecture on "the species of elephants, both living and fossil," Cuvier finally put an end to this way of thinking. Much as Charles Darwin is often credited with having come up with the theory of evolution—his real insight, of course, involved finding a mechanism for evolution—so Cuvier can be said to have theorized extinction.

Darwin's story has been recited (and re-recited) countless times by now. Entire books have been devoted to the few months he spent in Australia; to his mysterious and quite possibly psychosomatic illness; to the death of his oldest daughter; and to his decade-long study of barnacles. (This last subject is one that Darwin himself seems to have found tedious.) In 2009, when the two hundredth anniversary of Darwin's birth rolled around, the occasion was marked by scores of events, including an "evolution festival" in Vancouver, an uninterrupted reading of *On the Origin of Species* in Barcelona, and the construction of a massive Darwin doll for the Carnival parade in Recife. That same year, a full-length biopic, starring Jennifer Connelly as Darwin's wife (and first cousin), Emma, was released.

Cuvier, though, is very nearly forgotten. Many of his papers have still not been translated into English, and in studies of professional paleontology Cuvier is routinely slighted, even as he is acknowledged to be the founder of the discipline. Unless the situation changes dramatically, the two hundred fiftieth anniversary of his birth, in 2019, will pass without notice.

Darwin's work is inconceivable without Cuvier's discoveries. And yet Cuvier's obscurity is directly linked to Darwin's fame. Darwin's theory of extinction—that it was a routine side effect of evolution—contradicted Cuvier's, which held that species died out as a result of catastrophes, or, as he also put it, "revolutions on the surface of the earth." Darwin's view prevailed, Cuvier's was discredited, and for more than a century Cuvier was ignored. More recent discoveries, however, have tended to support the theories of Cuvier's that were most thoroughly vilified. Very occasionally, it turns out, the earth has indeed been wracked by catastrophe and, much as Cuvier imagined, "living organisms without number" have been their victims. This vindication of Cuvier would be of interest mainly to paleontologists and intellectual historians were it not for the fact that many scientists believe we are in the midst of such an event right now.

Since Cuvier's day, the National Museum of Natural History has grown into a sprawling institution with outposts all over France. Its main buildings, though, are still in Paris, on the site of the old royal gardens in the Fifth Arrondissement. Cuvier worked at the museum for most of his life and lived there, too, in a large stucco house that's been converted into office space. Next door to the house, there's a restaurant, and next to that a menagerie, where, on the day I visited, some wallabies were sunning themselves on the grass. Across the gardens, a large hall houses the museum's paleontology collection.

Pascal Tassy is a professor at the museum who specializes in proboscideans, the group that includes elephants and their lost cousins—mammoths, mastodons, and gomphotheres, to name just a few. He'd promised to show me the bones that Cuvier had examined when he came up with the theory of extinction. I found Tassy in his dimly lit office in the basement under the paleontology hall, sitting amid a mortuary's worth of old skulls. The walls of the office were decorated with covers from old *Tintin* comic books. Tassy told me he decided to become a paleontologist when he was seven, after reading a Tintin adventure about a dig.

We chatted about proboscideans for a while. "They're a fascinating group," he told me. "For instance, the trunk, which is a change of anatomy in the facial area that is truly extraordinary. It evolved separately five times. Two times—yes, that's surprising. But it happened *five* times, independently! We are forced to accept this by looking at the fossils." So far, Tassy said, some 170 proboscidean species have been identified, going back some 55 million years. "And this is far from complete, I am sure."

We headed upstairs to an annex attached to the back of the paleontology hall like a caboose. Tassy unlocked a small room crowded with metal cabinets. Just inside the door, partly wrapped in plastic, stood something resembling a hairy umbrella stand. This, he explained, was the leg of a woolly mammoth, which had been found, frozen and desiccated, on an island off Siberia. When I looked at it more closely, I could see that the skin of the leg had been stitched together like a moccasin. The hair was a very dark brown and seemed, even after more than ten thousand years, to be almost perfectly preserved.

Tassy opened one of the metal cabinets and placed its contents on a wooden table. These were some of the mastodon teeth that Cuvier had handled. The teeth had been found in the Ohio River Valley in 1739 by French soldiers, and, though they were there to fight a war, the soldiers had lugged the teeth down the Mississippi and put them on a boat to Paris.

"This is the *Mona Lisa* of paleontology," Tassy said, pointing to the largest of the group. "The beginning of everything. It's incredible, because Cuvier himself made the drawing of this tooth. So he looked at it very carefully." I picked it up in both hands. It was indeed a remarkable object. It was around 8 inches long and 4 across—about the size of a brick, and nearly as heavy. The cusps —four sets—were pointy, and the enamel was still largely intact. The roots, as thick as ropes, formed a solid mass the color of mahogany.

What particularly intrigued Cuvier about the mastodon teeth —and perplexed his predecessors—was that although they'd been found alongside a giant tusk, they didn't look anything like elephant teeth. Instead, they looked as though they could have belonged to an enormous human. (A mastodon molar that was sent to London in another eighteenth-century shipment was labeled "Tooth of a Giant.") In evolutionary terms, the explanation for this is simple: about 30 million years ago, the proboscidean line that would lead to mastodons split off from the line that would lead to elephants and also mammoths. The latter would eventually develop its more sophisticated teeth, which have ridges on the surface rather than cusps. (This arrangement is a lot tougher, and it allows elephants—and used to allow mammoths—to consume an unusually abrasive diet.)

Mastodons, meanwhile, retained their relatively primitive molars (as did humans) and just kept chomping away. Of course, as Tassy pointed out, the evolutionary perspective is precisely what Cuvier lacked, which in some ways makes his achievements that much more impressive.

"Sure, he made errors," Tassy said. "But his technical works—most of them are splendid. He was a real fantastic anatomist."

After we had examined the teeth a while longer, Tassy took me up to the paleontology hall. Just beyond the entrance, a giant femur, also sent from the Ohio River Valley to Paris, was displayed, mounted on a pedestal. It was as wide around as a fence post. French schoolchildren were streaming past us, yelling excitedly. Tassy had a large ring of keys, which he used to open various drawers underneath the glass display cases. He showed me a mammoth tooth that had been examined by Cuvier, and bits of various other extinct species that Cuvier had been the first to identify. Then we looked at one of the world's most famous fossils, known as the Maastricht animal—an enormous pointy jaw studded with shark-like teeth. In the eighteenth century, the Maastricht fossil was thought by some to belong to a strange crocodile and by others to be from a snaggletoothed whale. Cuvier attributed it, yet again correctly, to a marine reptile. (The creature was later dubbed a mosasaur.)

Around lunchtime, I walked Tassy back to his office and then wandered through the gardens to the restaurant next to Cuvier's old house. Because it seemed like the thing to do, I ordered the Menu Cuvier—your choice of entrée plus dessert. As I was working my way through the second course—a cream-filled tart—I began to feel uncomfortably full. I was reminded of a description I had read of the anatomist's anatomy. During the revolution, Cuvier

was thin. In the years he lived on the museum grounds, he grew
stouter and stouter until, toward the end of his life, he became
enormously fat.

With his lecture on "the species of elephants, both living and fossil," Cuvier had succeeded in establishing extinction as a fact. But
his most extravagant assertion—that there had existed a whole lost
world, filled with lost species—remained just that. If there had indeed been such a world, then it ought to be possible to find traces
of other extinct animals. So Cuvier set out to find them.

Paris in the 1790s was a fine place to be a paleontologist. The
hills to the north of the city were riddled with quarries that were
actively producing gypsum, the main ingredient of plaster of Paris.
(The capital grew so quickly over so many mines that cave-ins were
a major concern.) Not infrequently, quarriers came upon weird
bones, which were prized by collectors even though they had no
real idea what they were collecting. With the help of one such
enthusiast, Cuvier soon assembled the pieces of another extinct
animal, which he described as *l'animal moyen de Montmartre*—"the
medium-sized animal from Montmartre."

By 1800, four years after the elephant paper, Cuvier's fossil zoo
had expanded to include twenty-three species that he deemed to
be extinct. Among these were a pygmy hippopotamus, whose remains he found in a storeroom at the Paris museum; an elk with
enormous antlers, whose bones had been found in Ireland; and a
large bear—what now would be known as a cave bear—from Germany. The Montmartre animal had, by this point, divided, or multiplied, into six separate species. (Even today, little is known about
these species except that they were ungulates and lived some 30 to
40 million years ago.) "If so many lost species have been restored
in so little time, how many must be supposed to exist still in the
depths of the earth?" Cuvier asked.

Cuvier had a showman's flair and, long before the museum
employed public relations professionals, knew how to grab attention. ("He was a man who could have been a star on television
today," Tassy told me.) At one point, the gypsum quarries around
Paris yielded a fossil of a rabbit-size creature with a narrow body
and a squarish head. Cuvier hypothesized, based on the shape of
its teeth, that the fossil belonged to a marsupial. This was a bold
claim, as there were no known marsupials in the Old World. To

heighten the drama, Cuvier announced that he would put his identification to a public test. Marsupials have a distinctive pair of bones, now known as epipubic bones, that extend from their pelvis. Though these bones were not visible in the fossil as it was presented to Cuvier, he predicted that if he scratched around, the missing bones would be revealed. He invited Paris's scientific elite to gather and watch as he picked away at the fossil with a fine needle. Voilà, the bones appeared. (A cast of the marsupial fossil is on display in Paris in the paleontology hall, but the original is deemed too valuable to be exhibited and is kept in a special vault.)

Cuvier staged a similar bit of paleontological performance art during a trip to the Netherlands. In a museum in Haarlem, he examined a specimen that consisted of a large semicircular skull attached to part of a spinal column. The fossil, three feet long, had been discovered nearly a century earlier and had been attributed — rather curiously, given the shape of the head — to a human. (It had even been assigned a scientific name: *Homo diluvii testis,* or "man who was witness to the Flood.") To rebut this identification, Cuvier first found an ordinary salamander skeleton. Then, as Rudwick relates it, he began chipping away at the rock around the deluge man's spine. When he uncovered the fossil animal's forelimbs, they were, just as he had predicted, shaped like a salamander's. The creature was not an antediluvian human but something far weirder: a giant amphibian.

The more extinct species Cuvier turned up, the more the nature of the beasts seemed to change. Cave bears, giant sloths, even giant salamanders — all these bore some relation to species that were still alive. But what to make of a bizarre fossil that had been found in a limestone formation in Bavaria? Cuvier received an engraving of this fossil from one of his many correspondents. It showed a tangle of bones, including what looked to be extremely long arms, skinny fingers, and a narrow beak. The first naturalist to examine it speculated that its owner had been a sea animal and had used its elongated arms as paddles. Cuvier, on the basis of the engraving, determined — shockingly — that the animal was actually a flying reptile. He called it a *ptero-dactyle,* meaning "wing-fingered."

Cuvier's proof of extinction — of "a world previous to ours" — was a sensational event, and news of it soon spread across the Atlantic.

When a nearly complete giant skeleton was unearthed by some farm hands in Newburgh, New York, it was recognized as a find of great significance. Vice President Thomas Jefferson made several attempts to get his hands on the bones. He failed. But a friend, the artist Charles Willson Peale, who'd recently established the nation's first natural history museum, in Philadelphia, succeeded.

Peale, perhaps an even more accomplished showman than Cuvier, spent months fitting together the bones he acquired from Newburgh, fashioning the missing pieces out of wood and papier-mâché. He presented the skeleton to the public on Christmas Eve, 1801. To publicize the exhibition, Peale had his black servant, Moses Williams, don an Indian headdress and ride through the streets of Philadelphia on a white horse. The reconstructed beast stood eleven feet high at the shoulder and more than seventeen feet long from tusks to tail, a somewhat exaggerated size. Visitors were charged fifty cents—quite a considerable sum at the time —for a viewing. The beast, an American mastodon, at this point still lacked an agreed-upon name and was variously referred to as an *incognitum,* an Ohio animal, and, most confusing of all, a mammoth. It became America's first blockbuster exhibit and set off a wave of "mammoth fever." The town of Cheshire, Massachusetts, produced a 1,230-pound "mammoth cheese"; a Philadelphia baker produced a "mammoth bread"; and the newspapers reported on a "mammoth parsnip," a "mammoth peach tree," and a mammoth eater, who "swallowed 42 eggs in ten minutes." Peale also managed to piece together a second mastodon, out of additional bones found in Newburgh and a nearby town in the Hudson Valley. After a celebratory dinner held underneath the animal's capacious rib cage, he dispatched this second skeleton to Europe with two of his sons, Rembrandt and Rubens. The skeleton was exhibited for several months in London, during which time the younger Peales decided that the animal's tusks must have pointed downward, like a walrus's. Their plan was to take the skeleton on to Paris and sell it to Cuvier. But while they were in London, war broke out between Britain and France, making travel between the two countries impossible.

Cuvier finally gave the *mastodonte* its name in a paper published in Paris in 1806. The peculiar designation comes from the Greek, meaning "breast tooth"; the cusps on the animal's molars apparently reminded him of nipples.

Despite the ongoing hostilities between the British and the French, Cuvier managed to obtain detailed drawings of the skeleton that Peale's sons had taken to London, and these gave him a much better picture of the animal's anatomy. He realized that the mastodon was far more distant from modern elephants than the mammoth was, and assigned it to a new genus. (Today mastodons are given not only their own genus but their own family.) In addition to the American mastodon, Cuvier identified four other mastodon species, all "equally strange" to the earth today. Peale didn't learn of Cuvier's new name until 1809, and when he did he immediately seized on it. He wrote to Jefferson proposing a "christening" for the mastodon skeleton in his Philadelphia museum. Jefferson was lukewarm about the name Cuvier had come up with —it "may be as good as any other," he replied—and didn't deign to respond to the idea of a christening.

In 1812 Cuvier published a four-volume compendium of his work on fossil animals—*Recherches sur les Ossemens Fossiles de Quadrupèdes.* Before he began his "researches," there had been zero vertebrates classified as extinct. Thanks for the most part to his own efforts, there were now at least forty-nine.

As Cuvier's list grew, so did his renown. Few naturalists dared to announce their findings in public until he had vetted them. "Is not Cuvier the greatest poet of our century?" Balzac asked. "Our immortal naturalist has reconstructed worlds from a whitened bone; rebuilt, like Cadmus, cities from a tooth." Cuvier was honored by Napoleon and, once the Napoleonic Wars finally ended, was invited to Britain, where he was presented at court.

The English were eager converts to Cuvier's project. In the early years of the nineteenth century, fossil collecting became so popular among the upper classes that a whole new vocation sprang up. A "fossilist" was someone who made a living hunting up specimens for rich patrons. The year Cuvier published his *Recherches,* one such fossilist, a young woman named Mary Anning, discovered a particularly outlandish specimen. The creature's skull, found in the limestone cliffs of Dorset, was nearly four feet long, with a jaw shaped like a pair of needle-nose pliers. Its eye sockets, peculiarly large, were covered with bony plates.

The fossil ended up in London at the Egyptian Hall, a privately owned museum not unlike Peale's. It was put on exhibit as a fish and then as a relative of a platypus before being recognized as a new kind of reptile—an ichthyosaur, or "fish-lizard." A few years later, other specimens collected by Anning yielded pieces of another, even wilder creature, dubbed a plesiosaur, or "almost-lizard." Oxford's geology expert, the Reverend William Buckland, described the plesiosaur as having a lizardlike head joined to a neck "resembling the body of a Serpent," the "ribs of a Chameleon, and the paddles of a Whale." Apprised of the find, Cuvier found the account of the plesiosaur so outrageous that he questioned whether the specimen had been doctored. When Anning uncovered another, nearly complete plesiosaur fossil, Cuvier had to acknowledge that he'd been wrong. "One shouldn't anticipate anything more monstrous to emerge," he wrote to one of his British correspondents. During Cuvier's trip to England, he visited Oxford, where Buckland showed him yet another astonishing fossil—an enormous jaw with one curved tooth sticking up out of it like a scimitar. Cuvier recognized this animal, too, as some sort of lizard. A couple of decades later, the jaw was identified as belonging to a dinosaur.

The study of stratigraphy was in its infancy at this point, but it was already understood that different layers of rocks had been formed during different periods. The plesiosaur, the ichthyosaur, and the as yet unnamed dinosaur had all been found in limestone deposits that were attributed to what was then called the Secondary and is now known as the Mesozoic era. So, too, had the *ptero-dactyle* and the Maastricht animal. This pattern led Cuvier to another extraordinary insight about the history of life: it had a direction. Lost species whose remains could be found near the surface of the earth, like mastodons and cave bears, belonged to orders of creatures that were still alive. Digging back further, one found creatures, like the animals from Montmartre, that had no obvious modern counterparts. If one kept digging, mammals disappeared altogether from the fossil record. Eventually one reached not just a world previous to ours but a world previous to that, dominated by giant reptiles.

Cuvier's ideas about this history of life—that it was long, mutable, and full of fantastic creatures that no longer existed—would seem to have made him a natural advocate for evolution. But he opposed the concept of evolution, or *transformisme*, as it was known in Paris at the time, and he tried—generally, it seems, successfully —to humiliate any colleagues who advanced the theory. Curiously, it was the same skills that led him to discover extinction that made evolution appear to him preposterous, an affair as unlikely as alchemy.

As Cuvier liked to point out, he put his faith in anatomy; this was what had allowed him to distinguish the bones of a mammoth from those of an elephant and to recognize as a giant salamander what others took to be a man. At the heart of his understanding of anatomy was a notion that he termed "correlation of parts." By this he meant that the components of an animal all fit together and are optimally designed for its particular way of life; thus, a carnivore will have an intestinal system suited to digesting flesh. Its jaws will be "constructed for devouring prey; the claws, for seizing and tearing it; the teeth, for cutting and dividing its flesh; the entire system of its locomotive organs, for pursuing and catching it; its sense organs for detecting it from afar."

Conversely, an animal with hooves must be an herbivore, since it has "no means of seizing prey." It will have "teeth with a flat crown, to grind seeds and grasses," and a jaw capable of lateral motion. Were any one of these parts to be altered, the functional integrity of the whole would be destroyed. An animal that was born with, say, teeth or sense organs that were somehow different from its parents' would not be able to survive, let alone give rise to an entirely new kind of creature.

In Cuvier's day, the most prominent proponent of *transformisme* was his senior colleague at the National Museum of Natural History, Jean-Baptiste Lamarck. According to Lamarck, there was a force—the "power of life"—that pushed organisms to become increasingly complex. At the same time, animals and also plants often had to cope with changes in their environment. They did so by adjusting their habits; these new habits, in turn, produced physical modifications that were then passed down to their offspring. Birds that sought prey in lakes spread out their toes when they hit the water, and eventually developed webbed feet and became ducks. Moles, having moved underground, stopped using their sight, and so over generations their eyes became small and weak. Lamarck adamantly opposed Cuvier's idea of extinction; there was no process he could imagine that was capable of wiping an organism out entirely. (Interestingly, the only exception he entertained was humanity, which, Lamarck allowed, might be able to exterminate certain large and slow-to-reproduce animals.) What Cuvier interpreted as *espèces perdues* Lamarck claimed were simply those that had been most completely transformed.

The notion that animals could change their body types when convenient Cuvier found absurd. He lampooned the idea that "ducks by dint of diving became pikes; pikes by dint of happening upon dry land changed into ducks; hens searching for their food at the water's edge, and striving not to get their thighs wet, succeeded so well in elongating their legs that they became herons or storks." He discovered what was, to his mind at least, definitive proof against *transformisme* in a collection of mummies.

When Napoleon invaded Egypt, the French, as usual, seized whatever interested them. Among the crates of loot shipped back to Paris was an embalmed cat. Cuvier examined the mummy, looking for signs of transformation. He found none. The ancient Egyptian cat was, anatomically speaking, indistinguishable from a Parisian alley cat. This proved that species were fixed. Lamarck objected that the few thousand years that had elapsed since the Egyptian cat was embalmed represented "an infinitely small duration" relative to the vastness of time.

"I know that some naturalists rely a lot on the thousands of centuries that they pile up with a stroke of the pen," Cuvier responded dismissively. Eventually he was called upon to compose a eulogy for Lamarck, which he did very much in the spirit of burying rather than praising. Lamarck, according to Cuvier, was a fantasist. Like the "enchanted palaces of our old romances," his theories were built on "imaginary foundations," so that while they might "amuse the imagination of a poet," they could not "for a moment bear the examination of anyone who has dissected a hand, a viscus, or even a feather."

Having dismissed *transformisme*, Cuvier was left with a gaping hole. He had no account of how new organisms could appear, or any explanation for how the world could have come to be populated by different groups of animals at different times. This doesn't seem to have bothered him. His interest, after all, was not in the origin of species but in their demise.

The very first time he spoke about the subject, Cuvier intimated that he knew the driving force behind extinction, if not the exact mechanism. In his lecture on elephants, he proposed that the mastodon, the mammoth, and the *Megatherium* had all been wiped out "by some kind of catastrophe." Cuvier hesitated to speculate about the precise nature of the calamity—"It is not for us to involve ourselves in the vast field of conjectures that these questions open up"—but at that point he seems to have believed that one disaster would have sufficed.

Later, as his list of extinct species grew, his position changed. There had, he decided, been multiple cataclysms. "Life on earth has often been disturbed by terrible events," he wrote. "Living organisms without number have been the victims of these catastrophes."

Like his view of *transformisme,* Cuvier's belief in cataclysm fit with—indeed, could be said to follow from—his convictions about anatomy. Since animals were functional units, ideally suited to their circumstances, there was no reason, in the ordinary course of events, that they should die out. Not even the most devastating events known to occur in the contemporary world—volcanic eruptions, say, or forest fires—were sufficient to explain extinction; confronted with such changes, organisms simply moved on and survived. The changes that had caused extinctions must therefore have been of a much greater magnitude—so great that animals had been unable to cope with them. That such extreme events had never been observed by him or any other naturalist was another indication of nature's mutability: in the past, it had operated differently—more intensely and more savagely—than it did at present.

"The thread of operations is broken," Cuvier wrote. "Nature has changed course, and none of the agents she employs today would have been sufficient to produce her former works." Cuvier spent several years studying the rock formations around Paris—together with a mineralogist friend, he produced the first stratigraphic map of the Paris Basin—and here, too, he saw signs of cataclysmic change. The rocks showed that, at various points, the region had been submerged. The shifts from one environment to another —from marine to terrestrial or, at some points, from marine to freshwater—had, Cuvier decided, "not been slow at all"; rather, they had been brought about by those sudden "revolutions" on the surface of the earth. The latest of these revolutions must have occurred relatively recently, for traces of it were still everywhere apparent. This event, Cuvier believed, lay just beyond the edge of recorded history; he observed that many ancient myths and texts, including the Old Testament, allude to some sort of crisis—usually a deluge—that preceded the present order.

Cuvier's ideas about a globe wracked periodically by cataclysm proved to be nearly as influential as his original discoveries. His major essay on the subject, which was published in Paris in 1812, was almost immediately reprinted in English and exported to America. It also appeared in German, Swedish, Italian, and Russian. But a good deal was lost or, at least, misinterpreted in translation. Cuvier's essay was pointedly secular. He cited the Bible as merely one of many ancient texts, alongside the Hindu Vedas and the Shujing. This sort of ecumenism was unacceptable to the Anglican clergy who made up the faculty at institutions like Oxford, and when the essay was translated into English it was construed by Buckland and others as offering proof of Noah's flood.

By now the empirical grounds of Cuvier's theory have largely been disproved. The physical evidence that convinced him of a "revolution" just prior to recorded history (and that the English interpreted as proof of the Deluge) was in reality debris left behind by the last glaciation. The stratigraphy of the Paris Basin reflects not sudden "irruptions" of water but, rather, gradual changes in sea level and the effects of plate tectonics. On all these matters, Cuvier was, we now know, wrong.

Yet his wildest-sounding claims have turned out to be surprisingly accurate. Cataclysms happen. Nature does on occasion "change course," and at such moments it is as if the "thread of operations" has been broken. The contemporary term for these cataclysms is "mass extinctions," and the geological record suggests that in the past half-billion years, there have been five major ones and a dozen or more lesser ones. In the most severe of the so-called Big Five, at the end of the Permian period, some 250 million years ago, something like 90 percent of all species died off, and multicellular life came perilously close to being obliterated altogether. In the most recent, at the end of the Cretaceous, the dinosaurs were wiped out, along with the mosasaurs, the pterosaurs, the plesiosaurs, the ammonites, and two-thirds of all families of mammals, all in what, geologically speaking, amounted to an instant.

Meanwhile, as far as the American mastodon is concerned, Cuvier was to an almost uncanny extent correct. He decided that the beast had disappeared 5,000 or 6,000 years ago, in the same "revolution" that had killed off the mammoth and the *Megatherium*. Actually, the American mastodon vanished around 13,000 years ago, in a wave of disappearances that has become known as the megafauna extinction. This wave coincided with the spread of modern humans, and, increasingly, is understood to have been a result of it. Humans are now so rapidly transforming the planet—changing the atmosphere, altering the chemistry of the oceans, reshuffling the biosphere—that many scientists argue that we've entered a whole new geological epoch: the Anthropocene. In this sense, the crisis that Cuvier discerned just beyond the edge of recorded history was us.

II

The Geological Society of London, known to its members as the Geol Soc (pronounced "gee-ahl sock"), was founded in 1807, over dinner in a Covent Garden tavern. Geology was at that point a brand-new science, a circumstance reflected in the society's goals, which were to stimulate "zeal" for the discipline and to induce participants "to adopt one nomenclature." There followed long, often spirited debates on matters such as where to fix the borders of the Devonian period. "Though I don't much care for geology," one visitor to the society's early meetings noted, "I do like to see the fellows fight."

The Geol Soc is now headquartered in a stone mansion not far from Piccadilly Circus. On the outside the style of the mansion is Palladian; inside, it leans more toward midcentury public library. Much of the place is wrapped in plastic, owing to a construction project that never quite seems to reach completion. Near the reception desk, behind a green velvet curtain, hangs a copy of the first geological map of Britain, which was published in 1815 by William Smith. (Smith's British biographer has called the map "one of the classics of English science"; his American counterpart has pronounced it "the map that changed the world.") At the top of the stairs, there's a reading room with a brass chandelier, a few armchairs, some scuffed tables, and a broken coffee machine.

On a sunny morning not long ago, Jan Zalasiewicz, a stratigrapher and longtime society member, was sitting in the reading room, wishing the coffee machine were functional so that he could make a cup of tea. Zalasiewicz is a slight, almost elfin man with shaggy graying hair and narrow blue eyes. He had come down to London that morning from his home in Nottinghamshire to give a visitor a tour. His perspective on the Geol Soc, and on the city more generally, was, he had to admit, idiosyncratic.

"This building has never been considered as a rock before," he observed. "But it is just as much made of geology as anything you would find out in the field.

"Clearly, very few of these objects will survive Pompeii style," he went on, gesturing, with a faraway look in his eyes, toward the chairs, the tables, the magazine racks, and the coffee machine. "But they won't simply disappear. They'll break down into rubble, and the rubble will be washed away. But even the rubble that's been washed away will have its own character, its own signal." He swiveled to take in the windows (mostly silica) and the paneling (made of wood). "Potentially, everything here is fossilizable," he said.

Walter White–like, Zalasiewicz leads a double life. By day he's an expert on a group of ancient marine organisms known as graptolites. Zalasiewicz deeply admires graptolites, which thrived and diversified in the early Paleozoic, some 500 million years ago, only to be very nearly wiped out in a catastrophic extinction event. Present him with a fossilized graptolite and he can tell you at a glance which biozone of the Silurian period it belongs to.

In his off-hours, Zalasiewicz is a provocateur or, to be more British about it, "a scientific hooligan." He has more or less invented a new discipline, which might be called the stratigraphy of the future. It is based on a simple, if disturbing, premise: humans are so radically refashioning the planet—leveling so many forests, eliminating so many creatures that once occupied those forests, transporting so many other creatures around the globe, and burning through such vast quantities of fossil fuels to keep the whole enterprise going—that we may well end up producing a catastrophe comparable in scale to the one that laid waste to the graptolites. Already, Zalasiewicz is convinced, the geology of the planet has been permanently altered. The signal that will be left behind by our cities, our carbon emissions, and our potentially fossilizable detritus is strong enough, he maintains, that even a moderately competent stratigrapher, at a distance of 100 million years or so, should be able to tell that something extraordinary happened in what to us represents the present. "We have already left a record that is now indelible," he has written.

In recognition of the ways that, collectively, we are all world-changers, Zalasiewicz believes that an adjustment in nomenclature is called for. Officially, our epoch is the Holocene, but Zalasiewicz believes it would probably be more accurate to say that we have entered the Anthropocene. He is trying to persuade his colleagues to formally consider this new term. He hopes to bring the matter to a vote of the International Commission on Stratigraphy in 2016. If he has his way, every geology textbook in the world will instantly become obsolete.

The path led up a hill, across a stream, back across the stream, and past the carcass of a sheep, which looked deflated, like a lost balloon. The hill was bright green but treeless; generations of the sheep's relatives had kept anything from growing much above muzzle height. As far as I was concerned, it was raining. But in the Southern Uplands of Scotland, I was told, this counted only as a light drizzle, or smirr.

Zalasiewicz and I and two of his colleagues from the British Geological Survey had driven for more than five hours to get to the Uplands from the Survey's headquarters near Nottingham. We were hiking to a spot called Dob's Linn, where, according to an old ballad, the Devil himself was pushed over a precipice by a pious shepherd named Dob. By the time we reached the cliff, the smirr seemed to be smirring harder. There was a view over a waterfall, which crashed down into a narrow valley. A few yards farther up the path loomed a jagged outcropping of rock. It was striped vertically, like a referee's jersey, in bands of light and dark. Zalasiewicz set his rucksack down on the soggy ground and adjusted his red rain jacket. He pointed to one of the dark-colored stripes. "Bad things happened in here," he told me.

Much as Civil War buffs visit Gettysburg, stratigraphers are drawn to Dob's Linn. It's one of those rare places where, owing to an accident of plate tectonics, a major turning point in life's history is visible right on the surface of the earth. In this case, the event is the end-Ordovician extinction, which occurred some 440 million years ago. In addition to nearly knocking out the graptolites, it killed off something like 80 percent of the planet's species. ("Had the list of survivors been one jot different," Richard Fortey, a British paleontologist and a recent president of the Geol Soc, has observed, "then so would the world today.") Not coincidentally, Dob's Linn is also a great place to find graptolites.

To the naked eye, graptolite fossils look a bit like scratches and a bit like hieroglyphics. ("Graptolite" comes from the Greek, meaning "written rock"; the term was coined by Linnaeus, who dismissed graptolites as mineral encrustations trying to pass themselves off as the remnants of animals.) Viewed through a hand lens, they often prove to have lovely, evocative shapes; one species suggests a feather, another a lyre, a third the frond of a fern. Graptolites were colonial animals. Each one, known as a zooid, built itself a tiny, tubular shelter, known as a theca, that was attached to its neighbor's, like a row house. A single graptolite fossil thus represents a whole community, which drifted or, more probably, swam along as a single entity, feeding off even smaller plankton. Zalasiewicz lent me a hammer, and one of the graptolites I hacked out of the rock face had been preserved with peculiar clarity. It was shaped like a set of false eyelashes, but very small, as if for a Barbie. Zalasiewicz told me—doubtless exaggerating—that I had found a "museum-quality specimen." I pocketed it.

Graptolites had a habit—endearing from a stratigrapher's point of view—of speciating, spreading out, and dying off, all in relatively short order. Zalasiewicz likened them to Natasha, the tender heroine of *War and Peace*. They were, he told me, "delicate, nervous, and very sensitive to things around them." This makes them useful "index fossils"—successive species can be used to identify successive layers of rock.

Once Zalasiewicz showed me what to look for at Dob's Linn, I too could see that "bad things" happened here. The dark stripes were shale; in them, graptolites were plentiful and varied. This indicated that there was nothing alive to consume the animals once they'd died and sunk to the sea floor. Soon I'd collected so many that the pockets of my jacket were sagging. Many of the fossils were variations on the letter *V*, with two arms branching away from a central node. Some looked like zippers, others like wishbones. Still others had arms growing off their arms, like tiny trees.

The lighter stone—also shale—was barren, with barely a graptolite to be found in it. Paradoxically, this was a sign of a healthy ocean floor, with lots of scavengers living in the muck. The transition from one state to another—from gray stone to black, from no graptolites to many—appears to have occurred suddenly and, according to Zalasiewicz, *did* occur suddenly. "The change here from gray to black marks a tipping point, if you like, from a habitable sea floor to an uninhabitable one," Zalasiewicz said. "And one might have seen that in the span of a human lifetime." He described this transition as "Cuvierian."

Zalasiewicz's colleagues from the British Geological Survey, Dan Condon and Ian Millar, had come to Dob's Linn to collect samples from the various stripes. (Zalasiewicz also worked for many years at the BGS; he now teaches at the University of Leicester.) The samples, they hoped, would contain tiny crystals of zircon, which, after some complicated chemical manipulations, would allow them to date the layers of rock quite precisely. Millar, who grew up in Scotland, at first claimed to be undaunted by the smirr. But after a while even he admitted that it was pouring. Rivulets of mud were cascading down the face of the outcropping, compromising the samples. It was decided that we would have to come back the following day. The geologists packed up their gear and we squished back down the trail to the car. Zalasiewicz had made reservations at a bed-and-breakfast in the nearby town of Moffat. The town's attractions, I had read, included Britain's narrowest hotel and a bronze sheep.

The idea that the world can change suddenly and drastically—"in the span of a human lifetime"—is very old and, at the same time, very new. To the early members of the Geol Soc, the role of catastrophe in the earth's history was self-evident. These men—and they were, of course, all men—had read the great nineteenth-century French naturalist Georges Cuvier, who interpreted the fossil record as a chronicle of recurring tragedy. (When the Napoleonic Wars ended in 1815, Cuvier was made an honorary Geol Soc member.)

"Life on earth has often been disturbed by terrible events," Cuvier wrote. "Living organisms without number have been the victims of these catastrophes."

Cuvier's view of life was challenged by Charles Lyell, another of the nineteenth century's most influential naturalists. According to Lyell, who served as the Geol Soc's fourteenth president and also as its twenty-first, the earth was capable of changing only very gradually. The way to understand the distant past was to look at the present. Since no one had ever seen the kind of cataclysm that Cuvier invoked, it was unscientific or, to use Lyell's term, "unphilosophical," to imagine that such events took place. If it appeared from the fossil record that the world had changed abruptly, Lyell maintained, this just went to show how little the record was to be trusted.

Among the early converts to Lyell's view was Charles Darwin. In *On the Origin of Species,* Darwin acknowledged that there were points in the earth's history when it appeared that "whole families or orders" had suddenly been exterminated. But like Lyell, he took this as evidence that "wide intervals of time" were unaccounted for. Had the evidence of these intervals not been lost, it would have shown "much slow extermination." He wrote, "So profound is our ignorance, and so high our presumption, that we marvel when we hear of the extinction of an organic being; and as we do not see the cause, we invoke cataclysms to desolate the world!"

Such was Lyell and Darwin's influence that for more than a century, even as it became increasingly clear that "whole families or orders" had indeed at various points suddenly been eliminated, geologists eschewed any account of these episodes that might be construed as Cuvierian. This reluctance extended into the 1980s, when it was proposed that an asteroid plowing into the earth at the end of the Cretaceous period, 65 million years ago, was what had done in the dinosaurs, along with the plesiosaurs, the mosasaurs, the pterosaurs, the ammonites, most birds, and a significant proportion of mammals. The impact hypothesis was resisted until the 1990s, when the existence of a huge impact crater formed precisely at the end of the Cretaceous was confirmed. The crater lies off the Yucatán Peninsula, buried under half a mile of newer sediment.

While the discovery of the impact crater didn't exactly invalidate Lyell and Darwin's model, it revealed their dismissal of catastrophe to have been itself "unphilosophical." Life on earth *has* been "disturbed by terrible events," and "living organisms without number" have been their victims. What is sometimes called "neocatastrophism," but is mostly now just considered mainstream geology, holds that the world changes only very slowly, except when it doesn't.

As best as can be determined, the rate of change today is as fast as it's been at any time since the asteroid impact. This is why Zalasiewicz believes that the stratigraphers of the future should have a relatively easy time of it, even though who or what was responsible for the sudden alteration of the planet may not immediately be clear. At one point he mused, "It may take them a little while to sort out whether we were the drivers of this, or if the cats or the dogs or the sheep were."

After everyone had changed into dry clothes, we met in the sitting room of the B & B for tea. Zalasiewicz had brought along several papers he had recently published on graptolites. Settling back in their chairs, Condon and Millar rolled their eyes. Zalasiewicz ignored them, patiently explaining to me the import of his latest monograph, "Graptolites in British Stratigraphy," which ran to sixty-six pages and included illustrations of more than 650 species. In the monograph, the effects of the extinction event showed up more systematically, if also less vividly, than on the rain-slicked hillside. Until the end of the Ordovician, V-shaped graptolites were common. These included species like the *Dicranograptus ziczac*, whose tiny cups were arranged along arms that curled away and then toward each other, like tusks; and *Amphigraptus divergens*, which was shaped like a bat in flight. Only a handful of graptolite species survived the end-Ordovician extinction, which, it's now believed, was caused by the sudden glaciation of the supercontinent Gondwana. (No one is entirely sure what caused this glaciation.) Eventually the surviving graptolites diversified and repopulated the seas of the Silurian. But Silurian graptolites had a streamlined body plan, more like a stick than a set of branches. The V shape had been lost, never to reappear. Here, writ very, very small, was the fate of the dinosaurs, the pterosaurs, and the ammonites—a once highly successful form now relegated to oblivion.

That evening, when everyone had had enough of tea and graptolites, we went out to the pub on the ground floor of Britain's narrowest hotel, which is twenty feet across. After a pint or two, the conversation turned to another one of Zalasiewicz's favorite subjects: giant rats. Zalasiewicz pointed out that rats have followed humans to just about every corner of the globe, and it is his professional opinion that one day they will take over the earth.

"Some number will probably stay rat-size and rat-shaped," he told me. "But others may well shrink or expand. Particularly if there's been epidemic extinction, and ecospace opens up, rats may be best placed to take advantage of that. And we know that change in size can take place fairly quickly." I recalled once watching a rat drag a pizza crust along the tracks at an Upper West Side subway station. I imagined it waddling through a deserted tunnel, blown up to the size of a Doberman.

Though the connection might seem tenuous, Zalasiewicz's interest in giant rats represents a logical extension of his interest in graptolites. When he studies the Ordovician and the Silurian, he's trying to reconstruct the distant past on the basis of the fragmentary clues that remain—fossils, isotopes of carbon, layers of sedimentary rock. When he contemplates the future, he's trying to imagine what will remain of the present once the contemporary world has been reduced to fragments—fossils, isotopes of carbon, layers of sedimentary rock. One of the many aspects of the Anthropocene that he believes will leave a permanent mark is a reshuffling of the biosphere.

Often purposefully and just as often not, people have transported living things around the globe, importing the flora and fauna of Asia to the Americas and of the Americas to Europe and of Europe to Australia. Rats have consistently been in the vanguard of these movements, and they have left their bones scattered everywhere, including on islands so remote that humans never bothered to settle them. The Pacific rat, *Rattus exulans,* a native of Southeast Asia, traveled with Polynesian seafarers to, among many other places, Hawaii, Fiji, Tahiti, Tonga, Samoa, Easter Island, and New Zealand. Encountering few predators, stowaway *Rattus exulans* multiplied into what Richard Holdaway, a New Zealand paleontologist, has described as "a grey tide" that turned "everything edible into rat protein." (A recent study in the *Journal of Archaeological Science* concluded that it wasn't humans who deforested Easter Island; rather, it was the rats that came along for the ride and then bred unchecked. The native palms couldn't produce seeds fast enough to keep up with their appetite.)

When Europeans arrived in the Americas and then continued west to the islands that the Polynesians had settled, they brought

with them the even more adaptable Norway rat, *Rattus norvegicus.*
In many places, Norway rats, which are actually from China, out-
competed the earlier rat invaders and ravaged whatever bird and
reptile populations the Pacific rats had missed. Rats thus might
be said to have created their own "ecospace," which their progeny seem well positioned to dominate. The descendants of today's
rats, according to Zalasiewicz, will radiate out to fill the niches that
Rattus exulans and *Rattus norvegicus* helped empty. He imagines
the rats of the future evolving into new shapes and sizes—some
"smaller than shrews," others as large as elephants.

"We might," he has written in *The Earth After Us* (2008), "include
among them—for curiosity's sake and to keep our options open
—a species or two of large naked rodent, living in caves, shaping
rocks as primitive tools and wearing the skins of other mammals
that they have killed and eaten."

Meanwhile, whatever the future holds for rats, the extinction
event that they are helping to bring about will leave its own mark.
Many evolutionary lineages have recently come to an end; many,
many more are likely soon to follow. Extinction rates today are
hundreds of times higher—for some groups, such as amphibians
and freshwater mollusks, perhaps thousands, or even tens of thousands, of times higher—than they've been since mammals took
over the ecospace emptied by the dinosaurs. For reasons of geological history, the current extinction event is often referred to
as the "sixth extinction." (By this accounting, the event recorded
in the rocks at Dob's Linn is the first of the five major mass extinctions that have occurred since complex animal life evolved.)
Whether the sixth extinction will turn out to be anywhere near as
drastic as the first is impossible to know; nevertheless, it is likely
to appear in the fossil record as a turning point. Climate change
—itself a driver of extinction—will also leave behind geological
traces, as will deforestation, industrial pollution, and monoculture
farming.

Ultimately, most of our carbon emissions will end up in the
oceans; this will dramatically alter the chemistry of the water, turning it more acidic. Ocean acidification is associated with some
of the worst crises in biotic history, including what's known as
the end-Permian extinction—the third of the so-called Big Five
—which took place roughly 250 million years ago and killed off
something like 90 percent of the species on the planet.

"Oh, ocean acidification," Zalasicwicz said when we returned to Dob's Linn the following day. "That's the big nasty one that's coming down."

In recent years, a number of names have been proposed for the new age that humans have ushered in. The noted conservation biologist Michael Soulé has suggested that instead of the Cenozoic, we now live in the "Catastrophozoic" era. Michael Samways, an entomologist at South Africa's Stellenbosch University, has floated the term "Homogenocene." Daniel Pauly, a Canadian marine biologist, has recommended the "Myxocene," from the Greek word for "slime," and Andrew Revkin, an American journalist, has offered the "Anthrocene." (Most of these terms owe their origins, indirectly at least, to Lyell, who, back in the 1830s, coined the names Eocene, Miocene, and Pliocene.)

The word "Anthropocene" was put into circulation by Paul Crutzen, a Dutch chemist who, in 1995, shared a Nobel Prize for discovering the effects of ozone-depleting compounds. The importance of this discovery is difficult to overstate. Had it not been made—and had the chemicals continued to be widely used—the ozone "hole" that opens up every spring over Antarctica would have expanded until eventually it encircled the entire globe. One of Crutzen's fellow Nobelists reportedly came home from his lab one night and said to his wife, "The work is going well, but it looks like the end of the world."

Crutzen once told me that the word "Anthropocene" came to him while he was in a meeting. The meeting's chairman kept referring to the Holocene, the "wholly recent" epoch, which began at the conclusion of the last ice age, eleven and a half thousand years ago. According to the International Commission on Stratigraphy, or ICS, which maintains the official geological time scale, the Holocene continues to this day.

"'Let's stop it,'" Crutzen recalled blurting out. "'We are no longer in the Holocene; we are in the Anthropocene.' Well, it was quiet in the room for a while." At the next coffee break, the Anthropocene was the main topic of conversation. Someone came up to Crutzen and suggested that he patent the term.

Crutzen wrote up his idea in a short essay, titled "Geology of Mankind," which ran in the journal *Nature*. "It seems appropriate to assign the term 'Anthropocene' to the present, in many ways human-dominated, geological epoch," he observed. Among the many geologic-scale changes people have effected, Crutzen cited the following:

> Human activity has transformed between a third and a half of the land surface of the planet.
> Many of the world's major rivers have been dammed or diverted.
> Fertilizer plants produce more nitrogen than is fixed naturally by all terrestrial ecosystems.
> Humans use more than half of the world's readily accessible freshwater runoff.

Most significant, Crutzen noted, people have altered the composition of the atmosphere. Owing to a combination of fossil-fuel combustion and deforestation, the concentration of carbon dioxide in the air has risen by more than a third in the past two centuries, while the concentration of methane, an even more potent greenhouse gas, has more than doubled. Just a few more decades of emissions may bring atmospheric CO_2 to a level not seen since the mid-Miocene, 15 million years ago. A few decades after that, it could easily reach a level not seen since the Eocene, some 50 million years ago. During the Eocene, palm trees flourished in the Antarctic and alligators paddled around the British Isles.

"Because of these anthropogenic emissions," Crutzen wrote, the global climate is likely to "depart significantly from natural behavior for many millennia to come."

Crutzen published "Geology of Mankind" in 2002. Soon the Anthropocene began migrating into other scientific journals. "Global Analysis of River Systems: From Earth System Controls to Anthropocene Syndromes" was the title of a 2003 article in the journal *Philosophical Transactions of the Royal Society B.* "Soils and Sediments in the Anthropocene," ran the headline of a piece from 2004 in the *Journal of Soils and Sediments.*

Zalasiewicz noticed that most of those using the term were not trained in the fine points of stratigraphy, and he wondered how his colleagues felt about this. At the time, he was head of the Geol Soc's stratigraphy committee, and during a meeting one day he asked the members what they thought of the Anthropocene. Of the twenty-two stratigraphers present, twenty-one thought that the concept had merit.

"My response was it's a very interesting and powerful idea,"

Andy Gale, a professor at the University of Portsmouth, told me. "I felt it was worthwhile to pursue, because it's an important tool for making people think."

The group decided to approach the concept as a formal problem. Would the Anthropocene satisfy the stratigraphic criteria used for naming a new epoch? (To geologists, an epoch is a subdivision of a period, which, in turn, is a division of an era; the Holocene, for instance, is an epoch of the Quaternary, which is a period in the Cenozoic.) After a year's worth of study, the answer that the group arrived at was an unqualified yes. Among other things, the members observed in a paper summarizing their findings, the Anthropocene will be marked by a unique "biostratigraphic signal," a product of the current extinction event, on the one hand, and of the human propensity for redistributing life, on the other. This signal will be permanently inscribed, they wrote, "as future evolution will take place from surviving (and frequently anthropogenically relocated) stocks."

Or, as Zalasiewicz would have it, giant rats.

Just as in the early years of the Geol Soc, stratigraphers today spend a lot of time arguing about borders. A few years ago, after much heated discussion, members of the ICS voted to move the start of the Pleistocene epoch from about 1.8 million to about 2.6 million years ago. This decision was part of a broader, and even fiercer, debate about whether to do away with the Quaternary, the period that spans both the Pleistocene and the Holocene, and fold it into the Neogene. (The elimination of the Quaternary was vigorously—and ultimately successfully—resisted by Quaternary stratigraphers.)

The debate over the Anthropocene's borders is complicated by the fact that the geology of the epoch is, at this point, almost entirely prospective. The way stratigraphers usually define boundaries—once they've stopped arguing about them—is by choosing a particularly fossil-rich sequence of rocks to serve as a reference. These reference sequences are colloquially known as "golden spikes"—technically, they're called Global Boundary Stratotype Sections and Points, or GSSPs—and they're scattered around the world (though a disproportionate number are in Europe). The striped rocks at Dob's Linn have been designated the golden spike for the start of the Silurian period. For the base of the Carboniferous, the golden spike is near the town of Cabrières, in southern France, and for the start of the Triassic it's in the hills of Meishan, China. (The Chinese have tried to turn this last golden spike into a tourist destination, with a manicured park and a statue of a tooth from a once common eel-like creature known as a conodont.)

Since the rocks of the Anthropocene don't yet exist, it's impossible to choose an exemplary sequence of them. To stratigraphers, then, a key, but also rather vexing, question is what could serve instead of the traditional golden spike. In 2009 the ICS set up an Anthropocene Working Group to examine this and related issues; not surprisingly, Zalasiewicz was appointed chairman. At the time of our visit to London, he told me that he thought there were many possible ways that the start of the epoch could be designated. One would simply be to choose a date—1800, say, or 1950. This is how geological periods of the deep, pre-fossiliferous past are defined; what's known as the Neoproterozoic era, for example, is said to have begun precisely one billion years ago.

Another possibility would be to use nuclear fallout. The aboveground tests of the mid-twentieth century dispersed radioactive particles all around the globe. Some have half-lives of more than a thousand years; in a few cases, like uranium-236, the figure is in the tens of millions. To future geologists, the fallout will thus present a novel radioactive "spike" (assuming, that is, that the future does not hold a nuclear war). This sort of geochemical marker is used to define the end of the Cretaceous. The impact that occurred during the final seconds of the period left behind a thin layer of sediment containing anomalously high concentrations of the element iridium—the so-called iridium spike.

Yet another possibility is to use the world's subway systems, an idea that also has precedent in deep time. Geologists refer to the outlines of burrows that creatures left behind in the sediments as "trace fossils." The start of the Cambrian period, some 540 million years ago, is defined as the point when the first complex burrows appear; these left impressions in the rocks that resemble scattered grains of rice. (No one is sure what the animals that made the burrows looked like, as their bodies have not been preserved.) London's subway system, the world's oldest, will leave behind an enormous set of trace fossils, as will New York's and Seoul's and Paris's and Dubai's.

"All the great world cities have underground systems now,"

Mark Williams, a stratigrapher who teaches at the University of Leicester and is a member of the Anthropocene Working Group, noted. "They're extensive, they're fairly permanent from a geological perspective, and they're a very, very good indicator of the complexity that's come to characterize the twentieth and twenty-first centuries."

Williams told me that the response to the idea of formalizing the Anthropocene had "generally been very positive." (Just in the past few months, three new academic journals focusing on the Anthropocene have been launched.) But, as is to be expected from a group that can sustain a decade-long disagreement about the status of the Quaternary, there's still plenty of dissent. Some critics argue that humans have been altering the planet for thousands of years already, so why get all worked up about it now!'

"We can see that human interactions with the landscape are increasing," Philip Gibbard, a stratigrapher at Cambridge, told me. "No one disputes that. We build buildings. We build towns. We build roads. We drop plastic bags in the ocean. All that's absolutely true. But from a geological perspective—and I have to speak as a geologist, not as a generally interested person—I think what's happening now is just a logical continuation of something that began as human populations started to increase at the beginning of the Holocene.

"It is quite exciting to pursue this new idea," he added. "But I'm suspicious of it."

Other critics are skeptical of the idea for opposite reasons. They point out that human impacts on the planet are likely to become even more pronounced, and hence more stratigraphically significant, as time goes on. Thus, what's sometimes referred to in geological circles as the "event horizon" has not yet been reached.

For his part, Zalasiewicz is sympathetic to both lines of argument. Humans *have* been altering the planet for quite a while, though probably the impacts of the past were orders of magnitude more modest than they are today. And a few centuries from now the impacts of human activity may be orders of magnitude greater again. By the time people are through, Zalasiewicz told me, he wouldn't be surprised if the earth were rendered more or less unrecognizable. "One cannot exclude a P-T-type outcome," he observed, referring to the worst of the so-called Big Five, the end-

Permian, or Permo-Triassic, extinction. In the meantime, though, he said, "we have to work with what we've got."

This past summer, I went with Zalasiewicz on another collecting trip, this one to Wales. Zalasiewicz has a special fondness for the country. He wrote his dissertation on the stratigraphy of northern Wales, and while finishing his research he drove around in a decommissioned postal van and lived in a camper that had been used as a chicken coop. He wanted to show me a spot near the town of Ponterwyd where he thought there should perhaps be another golden spike—in this case, marking the base of the Aeronian Stage of the Silurian. We set out from the town of Keyworth, in Nottinghamshire, where Zalasiewicz lives with his wife and teenage son, and drove through the West Midlands. In its day the West Midlands was the industrial heart of Britain. Now the industry is mostly gone, and people struggle to find work. "About as scary an advertisement for the Anthropocene as you can imagine" is how Zalasiewicz described the region.

When we arrived at Ponterwyd, smirr was falling or, as the Welsh put it, piglaw. Again there were lots of sheep and green, sheep-shorn hills and rocks filled with fossils. Banging away at an outcropping, I soon found several graptolites. One, which Zalasiewicz identified as belonging to the species *Monograptus triangulatus,* looked like a tiny saw blade, with miniature triangular teeth. With characteristic tact, he told me that my specimen was "very lovely." I stuck it in my bag.

A few days later, I took the train back to London and then the Tube out to Heathrow, where I was spending the night at an airport hotel. Thanks to all the graptolites I'd gathered, my suitcase was overweight, and I decided that I was going to have to deaccession some of them. I took what seemed to be the least impressive examples and headed out through the lobby, only to realize that there was nowhere to go. The hotel faced a ten-foot wall, which was made of plywood and covered with billboard-size sheets of plastic printed with photographs of trees. The photos kept repeating, so that walking along was like getting lost in a dark monoculture. Beyond the plywood wall, there was a parking lot, and beyond that an access road. I figured that the parking lot would have to do. By this point I'd spent enough time with Zalasiewicz that the place appeared to me as a mosaic of human impacts. The lot was edged with a margin of dirt; this was filled with scraggly plants, many of them no doubt introduced species. Strewn among the weeds was the usual flotsam of travel: empty water bottles, crumpled candy wrappers, crushed soda cans, half-eaten packages of crisps. I recalled what Zalasiewicz had told me about aluminum, which is that until the late nineteenth century it did not exist on earth except in combination with other elements. So soda cans may provide yet another marker of our presence: the Dr Pepper spike.

It was a lovely evening. A half-moon hung in a purple sky crisscrossed by jet contrails. I took out my graptolites. Most I couldn't identify, but one, I thought, belonged to the species *Rhaphidograptus toernquisti*, which Zalasiewicz had described to me as among life's great success stories. *Rhaphidograptus toernquisti* managed to persist, unchanged, for some five million years. I placed my fossils in a little pile next to a discarded cigarette pack. Nearby I noticed a plastic pouch with the word TOXIC printed in block letters. The pouch was torn, and some ominously bright yellow powder was leaking out of it. I tried to imagine a geologist in the year A.D. 100,000,000 stumbling onto the site. It was hard for me to picture what he (or it) would look like, but I got a certain satisfaction thinking about how puzzled he would be when he came upon my Silurian graptolites nestled amid the wreckage of the Anthropocene.

JOSHUA LANG

Awakening

FROM *The Atlantic*

LINDA CAMPBELL WAS not quite four years old when her appendix burst, spilling its bacteria-rich contents throughout her abdomen. She was in severe pain, had a high fever, and wouldn't stop crying. Her parents, in a state of panic, brought her to the emergency room in Atlanta, where they lived. Knowing that Campbell's organs were beginning to fail and her heart was on the brink of shutting down, doctors rushed her into surgery.

Today removing an appendix leaves only a few droplet-size scars. But back then, in the 1960s, the procedure was much more involved. As Campbell recalls, an anesthesiologist told her to count backward from ten while he flooded her lungs with anesthetic ether gas, allowing a surgeon to slice into her torso, cut out her earthworm-size appendix, and drain her abdomen of infectious slop, leaving behind a lengthy longitudinal scar.

The operation was successful, but not long after Campbell returned home, her mother sensed that something was wrong. The calm, precocious girl who had gone into surgery was not the same one who emerged. Campbell began flinging food from her high chair. She suffered random episodes of uncontrollable vomiting. She threw violent temper tantrums during the day and had disturbing dreams at night. "They were about people being cut open, lots of blood, lots of violence," Campbell remembers. She refused to be alone but avoided anyone outside her immediate circle. Her parents took her to physicians and therapists. None could determine the cause of her distress. When she was in eighth grade, her parents pulled her from school for rehabilitation.

Over time Campbell's most severe symptoms subsided, and she learned how to cope with those that remained. She managed to move on, become an accountant, and start a family of her own, but she wasn't cured. Her nightmares continued, and nearly anything could trigger a panic attack: car horns, sudden bright lights, wearing tight-fitting pants or snug collars, even lying flat in a bed. She explored the possibility of posttraumatic stress disorder with her therapists but could not identify a triggering event. One clue that did eventually surface, though, hinted at a possibly traumatic experience. During a session with a hypnotherapist, Campbell remembered an image, accompanied by an acute feeling of fear, of a man looming over her.

Then, one fall afternoon in 2006, four decades after her symptoms began, Campbell met an anesthesiologist at a hypnotherapy workshop. Over lunch she found herself telling the anesthesiologist about her condition. She mentioned the appendectomy she'd had not long before everything changed.

The anesthesiologist was intrigued. He told her about a phenomenon that had sometimes accompanied early gas anesthetics, particularly ether, in which patients reacted to the gas by coughing and choking, as if they were suffocating.

The comment sparked something in Campbell. "I started having all these flashes," she remembers. "The flashes were me being on the table. The flashes were of the room. The flashes were of the bright lights over me." A man—the same one from her memory? —was there. At some point the room went black. "And then I got to the place where I was on the table, and I just remember feeling terror," she says. "That's all I remember. I don't see anything. I don't feel anything. It's absolute, abject terror. And the feeling that I am dying." At that moment, Campbell realized that something had happened to her during her appendectomy, something that changed her forever. After several years of investigation, she figured it out: she had woken up on the table.

This experience is called "intraoperative recall" or "anesthesia awareness," and it's more common than you might think. Although studies diverge, most experts estimate that for every thousand patients who undergo general anesthesia each year in the United States, one to two will experience awareness. Patients who awake hear surgeons' small talk, the swish and stretch of organs, the suctioning of blood; they feel the probing of fingers, the yanks and tugs on innards; they smell cauterized flesh and singed hair. But because one of the first steps of surgery is to tape patients' eyes shut, they can't see. And because another common step is to paralyze patients to prevent muscle twitching, they have no way to alert doctors that they are awake.

Many of these cases are benign: vague, hazy flashbacks. But up to 70 percent of patients who experience awareness suffer long-term psychological distress, including PTSD—a rate five times higher than that of soldiers returning from Iraq and Afghanistan. Campbell now understands that this is what happened to her, although she didn't believe it at first. "The whole idea of anesthesia awareness seemed over-the-top," she told me. "It took years to begin to say, 'I think this is what happened to me.'" She describes her memories of the surgery as being like those from a car accident: the moments before and after are clear, but the actual event is a shadowy blur of emotion. She searched online for people with similar experiences, found a coalition of victims, and eventually traveled up the East Coast to speak with some of them. They all shared a constellation of symptoms: nightmares, fear of confinement, the inability to lie flat (many sleep in chairs), and a sense of having died and returned to life. Campbell (whose name and certain other identifying details have been changed) struggles especially with the knowledge that there is no way for her to prove that she woke up and that many, if not most, people might not believe her. "Anesthesia awareness is an intrapersonal event," she says. "No one else sees it. No one else knows it. You're the only one."

In most cases of awareness, patients are awake but still dulled to pain. But that was not the case for Sherman Sizemore Jr., a Baptist minister and former coal miner who was seventy-three when he underwent an exploratory laparotomy in early 2006 to pinpoint the cause of recurring abdominal pain. In this type of procedure, surgeons methodically explore a patient's viscera for evidence of abnormalities. Although there are no official accounts of Sizemore's experience, his family maintained in a lawsuit that he was awake—and feeling pain—throughout the surgery. (The suit was settled in 2008.) He reportedly emerged from the operation behaving strangely. He was afraid to be left alone. He complained of being unable to breathe and claimed that people were trying to bury him alive. He refused to be around his grandchildren. He suffered from insomnia; when he could sleep, he had vivid nightmares.

The lawsuit claimed that Sizemore was tormented by doubt, wondering whether he had imagined the horrific pain. No one advised Sizemore to seek psychiatric help, his family alleged, and no one mentioned the fact that many patients who experience awareness suffer from PTSD. On February 2, 2006, two weeks after his surgery, Sizemore shot himself. He had no history of psychiatric illness.

Before the introduction of ether in the mid-nineteenth century, surgery was a rare and gruesome business. One of the most common operations was amputation. Surgeons used saws and knives to remove the offending appendage, and boiling oil and scalding irons to cauterize the wound. They resorted to a variety of methods, some more dangerous than others, to manage patients' pain. James Wardrop, a surgeon to the British royal family in the nineteenth century, wrote of a procedure called *deliquium animi*, in which he bled patients into quiescence. Others used alcohol, opiates, ice, tourniquets, or simple distraction.

The promise of painless surgery remained a preposterous idea in mainstream medicine until October 16, 1846. On that day, at the Harvard-affiliated Massachusetts General Hospital, a dentist named William Thomas Green Morton gave the first public demonstration of ether gas, administering it to a patient whose neck tumor was then removed by a surgeon. The event took place in a domed amphitheater now known as the "ether dome," and earned Harvard Medical School a truly international reputation. Oliver Wendell Holmes Sr., who coined the term anesthesia (from the Greek word *anaisthēsia*, meaning "lack of sensation"), rejoiced that "the fierce extremity of suffering has been steeped in the waters of forgetfulness, and the deepest furrow in the knotted brow of agony has been smoothed for ever." In 2007, when the *British Medical Journal* asked subscribers to name the most significant medical developments since 1840, anesthesia was among the top three, along with antibiotics and modern sanitation.

The miracle of anesthesia transcends pain. Painkillers—mainly opiates and alcohol—existed before ether, but they weren't sufficient to quell the nightmare of surgery. Ether accomplished something altogether different: it eliminated both experience and memory. When the drug wore off and patients woke up, their bodies stitched together and their minds intact, it was almost as though the intervening hours hadn't happened. The field that emerged from that historic moment in the ether dome was less concerned with the broad goal of curing disease than with a single task: the mastery of consciousness.

Anesthesia is often taken for granted in the daily routine of medicine today, both by health professionals and by the tens of millions of Americans who undergo surgery each year. Anesthesiologists are imbued with an almost heavenly power: with a mere push of their thumb on a clear plastic syringe, you go under. But in the past decade or so, several highly publicized cases, including Sherman Sizemore Jr.'s, have brought anesthesia awareness into the public forum. In 1998 a woman named Carol Weihrer, who claimed to have suffered awareness while having her eye removed, founded the Anesthesia Awareness Campaign, an advocacy group and resource for victims, and made the talk-show rounds. In 2007 the Hollywood thriller *Awake* intended, according to a producer, to "do to surgery what *Jaws* did to swimming in the ocean." Fearful of malpractice lawsuits, the profession grew defensive. The American Society of Anesthesiologists promised to find the cause of and solution for awareness. "Even one case is one too many," wrote the society's president in 2007.

This promise, however, is not so easily fulfilled. Despite 167 years of research, anesthesiologists still have little idea how their drugs unlock the mind. Which gears turn and unwind to produce oblivion? How do they turn back into place? These questions, as important as they are for preventing anesthesia awareness, are dwarfed by a central riddle that has puzzled scientists and philosophers—not to mention most mildly introspective people—for hundreds, if not thousands, of years: What does it mean to be conscious?

Doctors began investigating how anesthesia affects consciousness during the 1960s, shortly after the first reports of awareness. One South African researcher was especially curious about whether and how one might recall memories from a surgery. Perhaps a near-death experience? Pushing well beyond the limits of what would today be considered ethical, he collected ten volunteers undergoing dental surgery. The procedures went along as normal until, midway through, the room went silent and the medical staff reached for scripts.

"Just a moment!" the anesthesiologist would say. "I don't like the patient's color. Much too blue. His [or her] lips are very blue. I'm going to give a little more oxygen."

The anesthesiologist would then act out a medical emergency, rushing to the patient's bedside to ventilate his or her lungs, as if this action were necessary to save the patient's life. After several moments, the team would breathe a collective sigh of relief.

"There, that's better now," the anesthesiologist would affirm. "You can carry on with the operation."

A month later, the patients were hypnotized and asked to remember the day of the surgery. One female patient said she could hear someone talking in the operating theater.

"Who is it who's talking?" the interviewer asked.

"Dr. Viljoen," she said, referring to the anesthesiologist. "He's saying my color is gray."

"Yes?"

"He's going to give me some oxygen."

"What are his words?"

A long pause followed.

"He said that I will be all right now."

"Yes?"

"They're going to start again now. I can feel him bending close to me."

Of the ten volunteers, four remembered the words accurately; four retained vague memories; and two had no recollection of the surgery. The eight patients who did remember it displayed anxiety during the interview, many of them bursting from hypnosis, unable to continue. But when out of hypnosis, it was as though nothing had happened. They had no memory of the incident. The terror and anxiety seemed permanently buried in their subconscious.

This experiment revealed a fundamental problem for the study of awareness, the frequency of which can be measured only through reported accounts. For some victims, it can take weeks for memories to surface. For Linda Campbell, it took forty years. But what if no memory remains? Did awareness happen? Does it matter?

An anesthesiologist's job is surprisingly subjective. The same patient could be put under general anesthesia in a number of different ways, all to accomplish the same fundamental goal: to render him unconscious and immune to pain. Many methods also induce paralysis and prevent the formation of memory. Getting the patient under, and quickly, is almost always accomplished with propofol, a drug now famous for killing Michael Jackson. It is milky and viscous, almost like yogurt, in a fat syringe. When injected, it has a nearly instant hypnotic effect: blood pressure falls, heart rate increases, and breathing stops. (Anesthesiologists use additional drugs, as well as ventilation, to immediately correct for these effects.)

Other drugs in the anesthetic arsenal include fentanyl, which kills pain, and midazolam, which does little for pain but induces sleepiness, relieves anxiety, and interrupts memory formation. Rocuronium disconnects the brain from the muscles, creating a neuromuscular blockade, also known as paralysis. Sevoflurane is a multipurpose gaseous wonder, making it one of the most commonly used general anesthetics in the United States today—even though anesthesiologists are still relatively clueless as to how it produces unconsciousness. It crosses from the lungs into the blood, and from the blood to the brain, but . . . then what?

Other mysteries *have* been untangled. Redheads are known to feel pain especially acutely. This confused researchers until someone realized that the same genetic mutation that causes red hair also increases sensitivity to pain. One study found that redheaded patients require about 20 percent more general anesthesia than brunettes. Like redheads, children also require stronger anesthesia; their youthful livers clear drugs from the system much more quickly than adults' livers do. Patients with drug or alcohol problems, on the other hand, may be desensitized to anesthesia and require more—unless the patient is intoxicated at that moment, in which case less drug is needed.

After delivering the appropriate cocktail, anesthesiologists carefully monitor a patient's reactions. One way they do this is by tracking vital signs: blood pressure, heart rate, and temperature; fluid intake and urine output; oxygen saturation in arteries. They also observe muscles, pupils, breathing, and pallor, among many other indicators.

One organ, however, has remained stubbornly beyond their watch. Even though anesthesiologists are not entirely sure how their drugs work, they do know where they go: the brain. All changes in your vital signs are only the peripheral reverberations of anesthetic drugs' hammering on the soft mass inside your skull. Determining consciousness by measuring anything besides brain activity is like trying to decide whether a friend is angry by studying his or her facial expressions instead of asking directly, "Are you mad?"

In lamenting how little we know about the anesthetized brain, Gregory Crosby, a professor of anesthesiology at Harvard, wrote in the *New England Journal of Medicine* in 2011, "The astonishing thing is not that awareness occurs, but that it occurs so infrequently."

This ignorance gap seems almost absurd in the context of today's dazzling array of medical technologies. Doctors can parse your brain with innumerable X-ray slices and then collate them into a three-dimensional grayscale image in a process called computed tomography, or CT. They can send you into a tube where powerful magnets flip the spin of protons on water molecules in your brain; when the protons flip back into position, they emit radio waves, and from that information a computer can generate a comprehensive image known as an MRI (for magnetic resonance imaging). Positron emission tomography, or PET, scans provide detailed maps of metabolic activity. Yet we are in the Dark Ages when it comes to determining whether the brain is conscious or not? We can't figure out whether patients are awake, or what being awake even means?

Due to their hulking size, CT scanners and MRI machines are rarely, if ever, brought into an operating room. But other technologies are more mobile. For example, doctors can measure electrical activity in the brain using a machine that, with the help of a few electrodes attached to your scalp, generates what's known as an electroencephalogram (EEG) — essentially a snapshot of your brain waves. An EEG is printed in undulating longitudinal lines, like the scribbled outline of a mountain range: sometimes smooth and regular like the Appalachians, at other times rough and craggy like the Rockies, and in death or deep coma more like the Great Plains.

This technology is regularly used in sleep studies and to diagnose epilepsy and encephalitis, as well as to monitor the brain during certain specialized surgeries. But the problem with an EEG is interpreting it. The data come at a constant, unforgiving pace, with lines stacked one on top of another like a page of sheet music (high-density versions can have up to 256 lines). Before digitization, EEG printers disgorged paper at 30 millimeters per second, resulting in 324 meters of print for just three hours of surgery. And even today's machines provide data that are next to impossible to analyze on the fly, at least with any sort of detail or depth.

In 1985, when a twenty-three-year-old doctoral student named Nassib Chamoun first looked at the sheet music of an EEG, he saw a symphony—albeit one he could not yet read. Chamoun was then an electrical engineer doing a research fellowship at the Harvard School of Public Health, working on decoding the circuitry of the human heart. When an anesthesiologist he worked with argued that the brain was much more electrically interesting than the heart, Chamoun agreed to attend a demonstration of EEG, which was then a relatively new technology in the operating room, during a surgery at a Harvard hospital. The EEG printer was an old model, spilling reams of paper that piled near the head of the gurney. As the anesthesiologist injected the patient with drugs, the machine's pen danced wildly, ink splattering off the page. Chamoun was entranced by the complexity of the patterns he saw that day. He couldn't stop thinking about how to engineer these data into something that would be more useful for surgeons and anesthesiologists. He left his doctorate program and embarked on a twenty-five-year quest to decode the brain—and, ultimately, to quantify and measure consciousness.

As a child in Lebanon, Chamoun had been fascinated by taking things apart and putting them back together. During the 1970s, as ethnic tensions there boiled into civil war, Chamoun spent a lot of time cooped up at home when school was canceled or when it was unsafe for him to venture outside. The soldering iron and circuit board became his playground. Family members asked him to fix televisions, tape recorders, and radios. His parents gave him a microscope as a birthday present. He made his way to the United States for college, eager to expand his study beyond home electronics.

As it happened, the mid-1980s were an auspicious time for a young technologist with a promising idea. When Chamoun began working with EEG, he had early access to mainframe computers at Harvard and Boston University. More important, he had access to surgeries. He wheeled his digital EEG machine into Harvard operating rooms, fixed electrodes to patients, and recorded millions of data points. Then, using computers, he began to sift through the oceans of information, searching for a unifying pattern. Meanwhile, he was enlisting his old mentor, a Nobel Prize–winning Harvard professor, and courting venture-capital firms for seed money. The result was Aspect Medical Systems, a biotech firm he founded in 1987 with a singular goal: to build a monitor that anesthesiologists could use to discern their patients' level of awareness.

Chamoun turned out to have a pivotal ally in a family friend, Charlie Zraket, the CEO of a big defense contractor called Mitre. In the 1960s, mathematicians had developed a statistical method called bispectral analysis, which breaks down waveforms to find underlying patterns. This method was originally used for studying waves in the ocean, but Mitre applied it to voice-recognition software and later to sonar on war submarines and radar on airplanes. If bispectral analysis could be used to interpret patterns in ocean, radio, and sound waves, Zraket and Chamoun reasoned, why couldn't it be applied to brain waves?

Chamoun ended up banking everything on the belief that if he collected enough EEG data, he could hack the patterns using bispectral analysis. But by 1995, eight years in, the entire project was collapsing. Chamoun had gathered more than $18 million from investors, credit lines, and friends, and had spent it all, but still his algorithms could not reliably predict a patient's level of awareness. Just as he was confronting bankruptcy, he secured a $4 million investment from a well-known venture capitalist. This bought the time Chamoun needed. From that point, he and his team achieved a series of breakthroughs that caused them to fundamentally reframe the way they thought about consciousness. Chamoun had never believed that the brain was something with a simple on/off switch, but he had been looking for one master equation—a sort of electrical fingerprint of consciousness—that would connect all the dots. Only when he let go of the idea of a single equation did a new, more viable model come into view: consciousness as a spectrum of discrete phases that flowed one into the next, each marked by a different electrical fingerprint. Fully conscious to lightly sedated was one phase; lightly to moderately sedated was another; and so on. Once he realized this, Chamoun was able to identify at least five separate equations and arrange them in order, like snapping a series of photos and compiling them into a broad panorama shot.

The end product was a shoebox-size blue machine that used EEG data to rank a patient's level of awareness on an index of zero to 100, from coma to fully awake. Chamoun called it the Bispectral Index, or BIS, monitor. To use the BIS, all that anesthesiologists had to do was connect a pair of disposable electrode sensors to the machine, apply them firmly to a patient's forehead, and wait for a number to appear on the box's green-and-black digital display. They would then administer anesthesia and watch the number drop from a waking average of 97 to somewhere in the ideal "depth of anesthesia" range—between 40 and 60—at which point they could declare the patient ready for surgery.

The FDA cleared the BIS monitor in 1997. When *Time* interviewed Chamoun about the revolutionary device, he called it anesthesia's "Holy Grail." Two years later, Aspect Medical's quarterly revenue surpassed $8 million; the company soon went public. In 2000 Ernst & Young named Chamoun the Healthcare and Life Sciences Entrepreneur of the Year for the New England region.

Enthusiasm for the BIS monitor grew in 2004, when *The Lancet* published a groundbreaking study reporting that the device could reduce the incidence of anesthesia awareness by more than 80 percent. This nearly pushed the BIS into the realm of medical best practices. By July 2007 half of all American operating rooms had a BIS monitor. By 2010 the device had been used almost 40 million times worldwide. At his home in the suburbs of Boston, Chamoun has the 10-millionth sensor memorialized in a sealed plastic case.

The BIS monitor fundamentally changed the way scientists thought about consciousness. It compressed an enigmatic idea that had long mystified researchers into a medical indicator that could be quantified and measured, like blood pressure or body temperature. One effect of the accessibility of Chamoun's invention was that it was occasionally used outside the operating room, for purposes he had not foreseen. In a 2006 injunction involving a North Carolina death-row inmate named Willie Brown, a federal judge ruled that performing a lethal injection on a conscious prisoner could cause excessive pain. North Carolina requires prisons to anesthetize inmates before killing them, but the judge worried about the possibility of anesthesia awareness. Only when prison officials purchased a BIS monitor did he allow them to proceed with Brown's execution. So on April 21, 2006, attendants hooked Brown up to the monitor, injected him with a sedative, and watched his BIS value drop. At approximately two o'clock in the morning, once the number had fallen below 60, an attendant administered a lethal dose of pancuronium bromide and potassium chloride.

In the centuries before EEG and computers, the most active contemplators of consciousness were not doctors but philosophers. The seventeenth-century French thinker René Descartes proposed an influential theory that leaned on neuroanatomy as well as philosophical inference. He declared that the pineal gland, a pea-size glob just behind the thalamus, was the seat of consciousness, "the place in which all our thoughts are formed." But Descartes was a dualist: he believed that body and mind are separate and distinct. Within the physical matter of the pineal gland, he reasoned, something inexplicable must lie, something intangible —something that he identified as the soul.

This idea has been rejected by reductionist thinkers, who believe that consciousness is a scientific phenomenon that can be explained by the physiology of the brain. In an attempt to understand various sensory functions, a nineteenth-century cohort of reductionist biologists burned, cut, and excised lumps of the brain in rabbits, dogs, and monkeys, eventually pinpointing centers for hearing, vision, smell, touch, and memory. But even the most extreme experiments of the period failed to identify a center for consciousness. In 1892 a German scientist named Friedrich Goltz, who rejected this notion of cerebral localization and hypothesized instead that the brain operated as a cohesive unit, cut out the majority of a dog's cerebral cortex over the course of three operations. The animal managed to survive for eighteen months; it even remained active, walking its cage and curling up to sleep, and reacted to noises and light by flipping its ears and shutting its eyes. Yet other things had changed. The dog required assistance with eating, and its memory seemed to have been destroyed. "The condition was that of idiocy but not of unconsciousness," wrote one scientist.

Today's neuroscientists, most of whom are reductionists, have offered multiple hypotheses about where consciousness resides, from the anterior cingulate cortex, a region also associated with motivation, to some parts of the visual cortex, to the cytoskeleton structure of neurons. Some theories peg consciousness not to a particular part of the brain but to a particular process, such as the rhythmic activation of neurons between the thalamus and the cortex.

David Chalmers, an Australian philosopher who has written extensively about consciousness, would refer to this neurobiological hunt as the "easy problem." With enough time and money, scientists could ostensibly succeed in locating a consciousness center of the brain. But at that point, Chalmers argues, an even bigger mystery would still remain, one that he calls the "hard problem." Say you and a friend are looking at a sunset. Your body is processing a huge variety of sensory inputs: a spectrum of electromagnetic waves—red, orange, and yellow light—which focus on your retina; the vibrations of your friend's voice, which bounce along the bones of your inner ear and transform into a series of electrical signals that travel from neuron to neuron; memories of past sunsets, which spark a surge of dopamine in your mesolimbic pathway. These effects coalesce into one cohesive, indivisible experience of the sunset, one that differs from your friend's. The hard problem, or what the philosopher Joseph Levine called the "explanatory gap," is determining how physical and biological processes—all of them understood easily enough on their own—translate into the singular mystery of subjective experience. If this gap cannot be bridged, then consciousness must be informed by some sort of inexplicable, intangible element. And all of a sudden we are back to Descartes.

In 2004 a sixty-year-old man checked in for open gastric-bypass surgery and a gall-bladder removal at Virginia Mason Medical Center in Seattle. Simon, as I'll call him, stood five feet nine inches tall and weighed approximately three hundred pounds. In an open gastric bypass, the surgeon penetrates mounds of flesh and fat before finding the peritoneum, the glossy membrane that holds the abdominal cavity intact. Many surgeons use a space-age device called a Harmonic Scalpel, which cuts tissue while simultaneously blasting it with ultrasound waves to stop the bleeding. Once the surgeon uncovers the stomach and yards of folded, tubular intestines, she uses metal retractors to pull the skin apart and clear away slippery membranes, juicy organs, and fatty layers of tissue. Then to business: cut, suture, cut, suture, cauterize, cut.

No surgeon could have imagined a procedure of this magnitude 167 years ago, in the days before anesthesia. It would have been impossible to endure, both for the patient and the surgeon. Simon's anesthesiologist, Michael Mulroy, was particularly worried about him because of his hypotension and reliance on painkillers, both of which increased his risk of awareness. To make sure that Simon didn't drift into consciousness, Mulroy decided to use a BIS monitor.

Surgery records show that throughout the three-hour procedure, Simon's BIS value hovered between 37 and 51, well below the threshold for sedation. Mulroy had given Simon a relatively light dosage, reluctant to risk further deflating his patient's dangerously low blood pressure, but he took comfort in the fact that the BIS told him that Simon was unconscious and unaware.

After the surgery, in the postoperative recovery room, nurses asked Simon whether he was in pain. "Not now," he said, "but I was during surgery." Simon reported memories that began after intubation, including "unimaginable pain" and "the sensation that people were tearing at me." According to a clinical report, he heard voices around him and "wished he were dead," but when he tried to alert the surgical team, his body did not respond to his brain's commands.

The news of Simon's experience devastated Mulroy. He explained to his patient that he had used the BIS monitor and that it had confirmed Simon's unconscious state throughout the procedure. In the end, Mulroy says, all he could do was apologize and arrange for a psychiatrist. He hasn't seen Simon since, but he published the case in a 2005 issue of the journal *Anesthesia & Analgesia*. Mulroy felt that the BIS monitor had betrayed him; he might have done more to deepen Simon's sedation if the BIS had not reassured him that everything was fine.

Mulroy was one of the first to question the BIS, but his concern was soon echoed in other corners of the medical community. In 2008 the *New England Journal of Medicine* published a study comparing nearly 2,000 surgery patients at high risk of awareness: 967 patients were monitored by the BIS, and 974 via attention to changes in the amount of anesthetic gas they exhaled throughout a procedure. The author, a researcher at the Washington University School of Medicine in St. Louis named Michael Avidan, found that both groups of patients experienced awareness at similar rates. In other words, the BIS was no more effective than a much cheaper and more standard method. After questions were raised about his methodology, Avidan repeated the experiment with a broader sample and found the same thing. Chamoun's window to the brain, it turned out, was not especially enlightening.

Avidan, who seems to have a singular zeal for highlighting the BIS monitor's weaknesses, has also published a study showing that in many cases, two monitors on the same patient display different values. In a YouTube video, he applies BIS electrodes to a volunteer's forehead, cuts them with scissors, and waits a full forty seconds for the device's value to change. Today the BIS monitor has become the most controversial medical device in anesthesiology, if not all of surgery. Aspect's stock plummeted and the board of directors sold the company in 2009. Chamoun temporarily accepted a high-paying position at the new parent company, Covidien, but he resigned not long after. His heart wasn't in it anymore.

The BIS monitor is not obsolete: it may still be clinically useful, may still prevent some cases of awareness. "It's important to take into consideration the collective scientific evidence and clinical experience of millions of patients," says Chamoun. "The BIS can help reduce the risk of awareness, but it will not completely eliminate that risk."

Even after Avidan's studies, many anesthesiologists around the world still choose to rely on the BIS to guide them through surgery. But guarantee that a patient is unconscious? That it cannot do. Chamoun is an engineer: he was never interested in providing a holistic assessment of what it means to be conscious. For that, medicine had to hold out for someone who could see beyond the data—someone whose fascination with the mind was as much humanistic as scientific.

On a warm afternoon in Madison, Wisconsin, last spring, a psychiatrist was pointing an electromagnetic gun at my brain.

"Put your arm in your lap," he said.

I obeyed. My head was dressed in a sixty-electrode, high-density EEG-recording device. The doctor stood behind my chair, eyeing a digitized MRI of my brain and gliding the gun over my scalp until he found his target: my motor cortex.

"Relax."

I tried.

The gun clicked. My forehead muscles twitched. My arm leapt out of my lap, straight into the air, as if yanked by invisible puppet strings. "Do it again," I said.

This process is called transcranial magnet stimulation, or TMS. It is the key to a device that Giulio Tononi, one of the most talked-about figures in anesthesiology since Nassib Chamoun, hopes will provide a truly comprehensive assessment of consciousness. If successful, Tononi's device could reliably prevent anesthesia awareness. But his ambitions are much grander than that. Tononi is unraveling the mystery of consciousness: how it works, how to measure it, how to control it, and, possibly, how to create it.

At the heart of Tononi's work is his integrated-information theory, which is based on two distinct principles, as intuitive as they are scientific. First, consciousness is informative. Every waking moment of your life provides a nearly infinite reservoir of possible experiences, each one different from the next. Second, consciousness is integrated: you can't process this information in parts. When you see a red ball, you can't experience the color red separately from the shape of the ball. When you hear a word, you can't experience the sound of it separately from its meaning. According to this theory, a more conscious brain is both more informative (it has a deeper reservoir of experiences and stimuli) and more integrated (its perception of these experiences is more unified) than a less conscious one.

Compare the brain to New York City: just as cars navigate the city's neighborhoods via a patchwork of streets, bridges, tunnels, and highways, electrical signals traverse the brain via a meshwork of neurons. Tononi's theory predicts that in a fully conscious brain, traffic in one neighborhood will affect traffic in other neighborhoods, but that as consciousness fades—for instance, during sleep or anesthesia—this ripple effect will decrease or disappear.

In 2008, in one of several experiments demonstrating this effect, Tononi pulsed the brains of ten fully conscious subjects with his electromagnetic gun—the equivalent of, say, injecting a flood of new cars into SoHo. The traffic (the electromagnetic waves) rippled across Manhattan (the brain): things jammed up in Tribeca and Greenwich Village, even in Chelsea. Tononi's EEG electrodes captured ripples and reverberations that were different for every subject and for every region of the brain, patterns as complex and varied as the traffic in Manhattan on any given day.

Tononi then put the same subjects under anesthesia. Before he pulsed his gun again, the subjects' brain traffic seemed as busy as when they were conscious: cars still circulated in SoHo and Tribeca, in Greenwich Village and Chelsea. But the pulse had a drastically different effect: this time the traffic jam was confined to SoHo. No more ripples. "It's as if [the brain] has fragmented into pieces," Tononi told me. He published these findings in 2010 and also used them to file for a patent for "a method for assessing anesthetization."

I first encountered Giulio Tononi in 2011, at an American Society of Anesthesiologists conference, where he gave the final lecture. His voice—with an erudite Italian accent suitable for narrating the audio tour at the Sistine Chapel—echoed throughout the auditorium. His blond hair was parted in a zigzag across his head. He wore a brown suit with silver studs on the lapels, a white shirt, and a bolo tie. Here, speaking to a rapt audience of mostly American anesthesiologists, was an Italian neuroscientist dressed as if he were from Wyoming. "Anesthesia: the merciful annihilation of consciousness," Tononi said, a PowerPoint presentation projected behind him. "The one we devoutly wish for in the proper circumstances. Now, just like sleep takes consciousness away and gives it back, so does the anesthesiologist. Every day. He taketh and giveth."

On the next slide, Tononi projected Michelangelo's *The Creation of Eve*, which was captioned: *And the LORD God caused a deep sleep to fall upon Adam, and he slept: and he took one of his ribs, and closed up the flesh instead thereof.* "A quote from the very first surgical procedure that was done with anesthesia," Tononi said. "The operation was reasonably successful, it seems."

This tendency toward grandiloquence dates back at least to adolescence. As a teenager in Trento, a city in northern Italy, where his father was mayor, Tononi wrote a letter to Karl Popper, a famous European philosopher, asking him whether he should devote his life to studying consciousness.

Popper wrote back with encouragement and sent an inscribed copy of one of his books. Tononi considered approaching the subject through mathematics or philosophy, but ultimately decided that medicine would provide the best foundation. So he attended medical school, became a psychiatrist, and moved to New York for a fellowship under the physician Gerald Edelman. Although Edelman had won a Nobel Prize for his work in immunology, he had by that point pivoted to neuroscience. With Edelman, Tononi began publishing extensively on consciousness. He moved to Madison in 2001 and is now the Distinguished Chair in Consciousness Science at the University of Wisconsin.

When I visited Tononi in June to participate in one of his consciousness studies, he invited me to a dinner party at his home, fifteen minutes outside Madison. These dinners are legendary among his research fellows, many of them PhDs and MDs from Italy or Switzerland. I arrived at a luxurious log cabin replete with a tractor, an indoor swimming pool, and an outdoor pizza oven. Hanging over the oven was a wire sculpture in the shape of the Greek letter *phi*, which Tononi has chosen as the symbol for his consciousness theory and as the title of his latest book. The letter was also engraved on his bolo tie and commemorated on the license plate of his car.

He served a multicourse dinner featuring pasta made from scratch, pizzas cooked in the outdoor oven, a well-paired rosé, and absinthe. Midway through the second course, he asked each of his guests a question: whether we believed in free will, and why. I said that I didn't. I argued that if we are made of atoms based on physical laws, which form molecules ruled by chemical laws, which compose cells that abide by biological laws, how could there be free will? Tononi only smiled. If his theory of integration is correct, my logic is flawed, and free will can exist.

Tononi is to his neuroscientist peers as the eighteenth-century philosopher Immanuel Kant was to his empiricist counterpart David Hume. Like most modern neuroscientists, Hume saw only the "easy problem." He proposed that consciousness was nothing more and nothing less than the bundling of various bits of experiential knowledge or, as he called them, "perceptions." Using this logic, my physiological argument against free will could stand.

Kant, however, believed that the mind is more than an accumulation of experiences of the physical world. Like Descartes 150 years earlier and David Chalmers 200 years later, Kant focused on the "hard problem," making the logical argument that something beyond sensory inputs must account for the subjectivity of conscious experience—what Kant called "transcendental" consciousness. Tononi's theory hinges on a similar conception of conscious-

ness as something more than the sum of its experiential parts
—leaving room, then, for the possibility of free will.

The *amount* of integrated information in the brain—the quan-
tity of consciousness—is what Tononi calls phi. He believes he can
approximate it with a combination of his TMS-EEG technology
and mathematical models. Many well-known philosophers and
neuroscientists, however, remain skeptical. Chalmers has praised
Tononi for his bold attempt to quantify consciousness, but he
doesn't think Tononi has come any closer to solving the hard
problem. And even Tononi admits that, in scientific-research time,
his theory is still in its infancy. What Tononi has made progress on
is neither the easy nor the hard problem: it's the practical problem. He is currently developing a machine that has the potential
to end anesthesia awareness once and for all. Like the BIS monitor, the device would provide a numerical assessment of a patient's
awareness and would be simple and compact enough to become
a regular fixture in operating rooms. Unlike the BIS monitor, it
would also be relevant outside the operating room. Whereas the
BIS is rooted in data specific to surgery, Tononi has developed a
comprehensive theory of consciousness that could, with appropriate technological tweaks, be applied in any number of medical,
scientific, or social settings.

My experience with the electrode cap and TMS gun in Tononi's
lab offers a rough guide to how his awareness monitor might work.
First, an anesthesiologist would dress your head with electrodes,
which would transmit EEG data to a processor. After putting you
under, she would monitor the drugs' effects by using a paddle attached to a high-voltage generator to repeatedly blast your brain
with electromagnetic waves. The EEG processor would monitor
your brain's reaction to each blast, calculating the complexity of
patterns and the degree of integration and ultimately displaying
a numerical phi value—your level of consciousness. Returning to
the New York metaphor: if the traffic jam stayed in SoHo, the machine would display a low value, and the anesthesiologist could
relax; if it spread to other neighborhoods, the machine would display a high value, and she might want to administer more drugs.

While this device is still millions of dollars and many trial hours
away from implementation, Tononi and his Italian colleague Mar-
cello Massimini have tested and validated an approximation of phi

in multiple settings, and are preparing to publish their findings. The method has already been used by some clinics in Europe—not on anesthetized patients but on vegetative ones. What if Terri Schiavo's family had been able to ascertain that she was, in fact, completely unconscious, more so than she would have been even under heavy anesthesia? These clinics are calculating phi to assess whether comatose patients experience consciousness, and if so, how much. Of course, this application has troubling risks; a flaw in Tononi's theory could lead families to turn off life support for a still-conscious person. But if the theory holds up—if Tononi has successfully managed to quantify consciousness—it could deliver these families from uncertainty. It could also change the way we think about animal rights, upend the abortion debate—possibly even revolutionize the way we think about artificial intelligence. But before any of that, it would fulfill the promise first offered in the ether dome more than a century and a half ago: an end to the nightmare of waking surgery.

In his recently published book, *Phi*, Tononi narrates a literary tour of his theory of consciousness through a fictionalized protagonist: Galileo. In one of the last chapters, Galileo encounters a diabolical machine that surgically manipulates the brain to produce pure sensations of pain. Tononi calls it "the only real and eternal hell." The creator of the machine asks: "What is the perfect pain? Can pain be made to last forever? Did pain exist, if it leaves no memory? And is there something worse than pain itself?"

For George Wilson, a Scottish chemist who had his foot amputated in 1843, before the dissemination of anesthesia, pain gave way to something seemingly beyond physical sensation, something articulable only in spiritual, nearly existential terms. Wilson described his experience in a letter several years after his surgery:

> Of the agony it occasioned I will say nothing. Suffering so great as I underwent cannot be expressed in words, and fortunately cannot be recalled. The particular pangs are now forgotten, but the blank whirlwind of emotion, the horror of great darkness, and the sense of desertion by God and man, bordering close upon despair, which swept through my mind and overwhelmed my heart, I can never forget, however gladly I would do so.

While subduing consciousness is the most urgent aspect of Tononi's work, he is especially animated when discussing consciousness in its fullest, brightest state. In his office in Madison, he described a hypothetical device called a "qualiascope" that could visualize consciousness the same way telescopes visualize light waves, or thermal goggles visualize heat. The more integrated the information—that is, the more conscious the brain—the brighter the qualiascope would glow. Using the device in an operating room, you would watch a patient's consciousness fade to a dull pulse. If he woke up midoperation, you might see a flicker.

But if you turned your gaze away from the operating room, you would gain an astonishing perspective on the universe. "The galaxy would look like dust," Tononi told me. "Within this empty, dusty universe, there would be true stars. And guess what? These stars would be every living consciousness. It's really true. It's not just a poetic image. The big things, like the sun, would be nothing compared to what we have."

MARYN McKENNA

Imagining the Post-Antibiotics Future

FROM *Medium*

A FEW YEARS AGO, I started looking online to fill in chapters of my family history that no one had ever spoken of. I registered on Ancestry.com, plugged in the little I knew, and soon was found by a cousin whom I had not known existed, the granddaughter of my grandfather's older sister. We started exchanging documents: a copy of a birth certificate, a photo from an old wedding album. After a few months, she sent me something disturbing.

It was a black-and-white scan of an article clipped from the long-gone *Argus* of Rockaway Beach, New York. In the scan, the type was faded and there were ragged gaps where the soft newsprint had worn through. The clipping must have been folded and carried around for a long time before it was pasted back together and put away. The article was about my great-uncle, the younger brother of my cousin's grandmother and my grandfather.

In a family that never talked much about the past, he had been discussed even less than the rest. I knew he had been a fireman in New York City and had died young, and that his death scarred his family with a grief they never recovered from. I knew that my father, a small child when his uncle died, was thought to resemble him. I also knew that when my father made his Catholic confirmation a few years afterward, he chose as his spiritual guardian the saint that his uncle had been named for: Saint Joseph, the patron of a good death.

I had always heard Joe had been injured at work: not burned, but bruised and cut when a heavy brass hose nozzle fell on him. The article revealed what happened next. Through one of the scrapes, an infection set in. After a few days, he developed an ache in one shoulder; two days later, a fever. His wife and the neighborhood doctor struggled for two weeks to take care of him, then flagged down a taxi and drove him 15 miles to the hospital in my grandparents' town. He was there one more week, shaking with chills and muttering through hallucinations, and then sinking into a coma as his organs failed. Desperate to save his life, the men from his firehouse lined up to give blood. Nothing worked. He was thirty when he died, in March 1938.

The date is important. Five years after my great-uncle's death, penicillin changed medicine forever. Infections that had been death sentences—from battlefield wounds, industrial accidents, childbirth—suddenly could be cured in a few days. So when I first read the story of his death, it lit up for me what life must have been like before antibiotics started saving us.

Lately, though, I read it differently. In Joe's story I see what life might become if we did not have antibiotics anymore.

Predictions that we might sacrifice the antibiotic miracle have been around almost as long as the drugs themselves. Penicillin was first discovered in 1928 and battlefield casualties got the first non-experimental doses in 1943, quickly saving soldiers who had been close to death. But just two years later, the drug's discoverer Sir Alexander Fleming warned that its benefit might not last. Accepting the 1945 Nobel Prize in Medicine, he said: "It is not difficult to make microbes resistant to penicillin in the laboratory by exposing them to concentrations not sufficient to kill them . . . There is the danger that the ignorant man may easily underdose himself and by exposing his microbes to non-lethal quantities of the drug make them resistant."

As a biologist, Fleming knew that evolution was inevitable: sooner or later, bacteria would develop defenses against the compounds the nascent pharmaceutical industry was aiming at them. But what worried him was the possibility that misuse would speed the process up. Every inappropriate prescription and insufficient dose given in medicine would kill weak bacteria but let the strong

survive. (As would the micro-dose "growth promoters" given in agriculture, which were invented a few years after Fleming spoke.)
Bacteria can produce another generation in as little as twenty minutes; with tens of thousands of generations a year working out survival strategies, the organisms would soon overwhelm the potent
new drugs.

Fleming's prediction was correct. Penicillin-resistant staph
emerged in 1940 while the drug was still being given to only a few
patients. Tetracycline was introduced in 1950, and tetracycline-re-
sistant *Shigella* emerged in 1959; erythromycin came on the market
in 1953, and erythromycin-resistant strep appeared in 1968. As antibiotics became more affordable and their use increased, bacteria
developed defenses more quickly. Methicillin arrived in 1960 and
methicillin resistance in 1962; levofloxacin in 1996 and the first
resistant cases the same year; linezolid in 2000 and resistance to
it in 2001; daptomycin in 2003 and the first signs of resistance in
2004.

With antibiotics losing usefulness so quickly—and thus not making back the estimated $1 billion per drug it costs to create them
—the pharmaceutical industry lost enthusiasm for making more.
In 2004, there were only five new antibiotics in development, compared to more than five hundred chronic-disease drugs for which
resistance is not an issue—and which, unlike antibiotics, are taken
for years, not days. Since then, resistant bugs have grown more numerous and, by sharing DNA with each other, have become even
tougher to treat with the few drugs that remain. In 2009 and again
this year, researchers in Europe and the United States sounded
the alarm over an ominous form of resistance known as CRE, for
which only one antibiotic still works.

Health authorities have struggled to convince the public that
this is a crisis. In September, Dr. Thomas Frieden, the director
of the U.S. Centers for Disease Control and Prevention, issued a
blunt warning: "If we're not careful, we will soon be in a post-antibiotic era. For some patients and some microbes, we are already
there." The chief medical officer of the United Kingdom, Dame
Sally Davies—who calls antibiotic resistance as serious a threat as
terrorism—recently published a book in which she imagines what
might come next. She sketches a world where infection is so dangerous that anyone with even minor symptoms would be locked in
confinement until they recover or die. It is a dark vision, meant to

disturb. But it may actually underplay what the loss of antibiotics would mean.

In 2009, three New York physicians cared for a sixty-seven-year-old man who had major surgery and then picked up a hospital infection that was "pan-resistant"—that is, responsive to no antibiotics at all. He died fourteen days later. When his doctors related his case in a medical journal months afterward, they still sounded stunned. "It is a rarity for a physician in the developed world to have a patient die of an overwhelming infection for which there are no therapeutic options," they said, calling the man's death "the first instance in our clinical experience in which we had no effective treatment to offer."

They are not the only doctors to endure that lack of options. Dr. Brad Spellberg of UCLA's David Geffen School of Medicine became so enraged by the ineffectiveness of antibiotics that he wrote a book about it. "Sitting with a family, trying to explain that you have nothing left to treat their dying relative—that leaves an indelible mark on you," he says. "This is not cancer; it's infectious disease, treatable for decades."

As grim as they are, in-hospital deaths from resistant infections are easy to rationalize: perhaps these people were just old, already ill, different somehow from the rest of us. But deaths like this are changing medicine. To protect their own facilities, hospitals already flag incoming patients who might carry untreatable bacteria. Most of those patients come from nursing homes and "long-term acute care" (an intensive-care alternative where someone who needs a ventilator for weeks or months might stay). So many patients in those institutions carry highly resistant bacteria that hospital workers isolate them when they arrive and fret about the danger they pose to others. As infections become yet more dangerous, the health care industry will be even less willing to take such risks.

Those calculations of risk extend far beyond admitting possibly contaminated patients from a nursing home. Without the protection offered by antibiotics, entire categories of medical practice would be rethought.

Many treatments require suppressing the immune system, to help destroy cancer or to keep a transplanted organ viable. That suppression makes people unusually vulnerable to infection. Antibiotics reduce the threat; without them, chemotherapy or radiation treatment would be as dangerous as the cancers they seek to cure. Dr. Michael Bell, who leads an infection-prevention division at the CDC, told me: "We deal with that risk now by loading people up with broad-spectrum antibiotics, sometimes for weeks at a stretch. But if you can't do that, the decision to treat somebody takes on a different ethical tone. Similarly with transplantation. And severe burns are hugely susceptible to infection. Burn units would have a very, very difficult task keeping people alive."

Doctors routinely perform procedures that carry an extraordinary infection risk unless antibiotics are used. Chief among them: any treatment that requires the construction of portals into the bloodstream and gives bacteria a direct route to the heart or brain. That rules out intensive-care medicine, with its ventilators, catheters, and ports—but also something as prosaic as kidney dialysis, which mechanically filters the blood.

Next to go: surgery, especially on sites that harbor large populations of bacteria such as the intestines and the urinary tract. Those bacteria are benign in their regular homes in the body, but introduce them into the blood, as surgery can, and infections are practically guaranteed. And then implantable devices, because bacteria can form sticky films of infection on the devices' surfaces that can be broken down only by antibiotics.

Dr. Donald Fry, a member of the American College of Surgeons, who finished medical school in 1972, says: "In my professional life, it has been breathtaking to watch what can be done with synthetic prosthetic materials: joints, vessels, heart valves. But in these operations, infection is a catastrophe." British health economists with similar concerns recently calculated the costs of antibiotic resistance. To examine how it would affect surgery, they picked hip replacements, a common procedure in once-athletic baby boomers. They estimated that without antibiotics, one out of every six recipients of new hip joints would die.

Antibiotics are administered prophylactically before operations as major as open-heart surgery and as routine as cesarean sections and prostate biopsies. Without the drugs, the risks posed by those operations, and the likelihood that physicians would perform them, will change.

"In our current malpractice environment, is a doctor going to want to do a bone marrow transplant, knowing there's a very high rate of infection that you won't be able to treat?" asks Dr. Louis Rice, chair of the department of medicine at Brown University's medical school. "Plus, right now health care is a reasonably free-market, fee-for-service system; people are interested in doing procedures because they make money. But five or ten years from now, we'll probably be in an environment where we get a flat sum of money to take care of patients. And we may decide that some of these procedures aren't worth the risk."

Medical procedures may involve a high risk of infections, but our everyday lives are pretty risky too. One of the first people to receive penicillin experimentally was a British policeman, Albert Alexander. He was so riddled with infection that his scalp oozed pus and one eye had to be removed. The source of his illness: scratching his face on a rosebush. (There was so little penicillin available that, though Alexander rallied at first, the drug ran out and he died.)

Before antibiotics, five women died out of every one thousand who gave birth. One out of nine people who got a skin infection died, even from something as simple as a scrape or an insect bite. Three out of ten people who contracted pneumonia died from it. Ear infections caused deafness; sore throats were followed by heart failure. In a post-antibiotic era, would you mess around with power tools? Let your kid climb a tree? Have another child?

"Right now, if you want to be a sharp-looking hipster and get a tattoo, you're not putting your life on the line," says the CDC's Bell. "Botox injections, liposuction, those become possibly life-threatening. Even driving to work: we rely on antibiotics to make a major accident something we can get through, as opposed to a death sentence."

Bell's prediction is a hypothesis for now—but infections that resist even powerful antibiotics have already entered everyday life. Dozens of college and professional athletes, most recently Lawrence Tynes of the Tampa Bay Buccaneers, have lost playing time or entire seasons to infections with drug-resistant staph, MRSA. Girls who sought permanent-makeup tattoos have lost their eyebrows after getting infections. Last year three members of a Maryland family—an elderly woman and two adult children—died of resistant pneumonia that took hold after simple cases of flu.

At UCLA, Spellberg treated a woman with what appeared to be

an everyday urinary tract infection—except that it was not quelled by the first round of antibiotics, or the second. By the time he saw her, she was in septic shock, and the infection had destroyed the bones in her spine. A last-ditch course of the only remaining antibiotic saved her life, but she lost the use of her legs. "This is what we're in danger of," he says. "People who are living normal lives who develop almost untreatable infections."

In 2009 Tom Dukes—a fifty-four-year-old inline skater and bodybuilder—developed diverticulosis, a common problem in which pouches develop in the wall of the intestine. He was coping with it, watching his diet and monitoring himself for symptoms, when searing cramps doubled him over and sent him to urgent care. One of the thin-walled pouches had torn open and dumped gut bacteria into his abdomen—but for reasons no one could explain, what should have been normal *E. coli* were instead highly drug-resistant. Doctors excised 8 inches of his colon in emergency surgery. Over several months, Dukes recovered with the aid of last-resort antibiotics, delivered intravenously. For years afterward, he was exhausted and in pain. "I was living my life, a really healthy life," he says. "It never dawned on me that this could happen."

Dukes believes, though he has no evidence, that the bacteria in his gut became drug-resistant because he ate meat from animals raised with routine antibiotic use. That would not be difficult: most meat in the United States is grown that way. To varying degrees, depending on their size and age, cattle, pigs, and chickens —and, in other countries, fish and shrimp—receive regular doses to speed their growth, increase their weight, and protect them from disease. Out of all the antibiotics sold in the United States each year, 80 percent by weight are used in agriculture, primarily to fatten animals and protect them from the conditions in which they are raised.

A growing body of scientific research links antibiotic use in animals to the emergence of antibiotic-resistant bacteria: in the animals' own guts, in the manure that farmers use on crops or store on their land, and in human illnesses as well. Resistant bacteria move from animals to humans in groundwater and dust, on flies, and via the meat those animals get turned into.

An annual survey of retail meat conducted by the Food and Drug Administration—part of a larger project involving the CDC and the U.S. Department of Agriculture that examines animals, meat, and human illness—finds resistant organisms every year. In its 2011 report, published last February, the FDA found (among many other results) that 65 percent of chicken breasts and 44 percent of ground beef carried bacteria resistant to tetracycline, and 11 percent of pork chops carried bacteria resistant to five classes of drugs. Meat transports those bacteria into your kitchen, if you do not handle it very carefully, and into your body if it is not thoroughly cooked—and resistant infections result.

Researchers and activists have tried for decades to get the FDA to rein in farm overuse of antibiotics, mostly without success. The agency attempted in the 1970s to control agricultural use by revoking authorization for penicillin and tetracycline to be used as "growth promoters," but that effort never moved forward. Agriculture and the veterinary pharmaceutical industry pushed back, alleging that agricultural antibiotics have no demonstrable effect on human health.

Few, though, have asked what multi-drug–resistant bacteria might mean for farm animals. Yet a post-antibiotic era imperils agriculture as much as it does medicine. In addition to growth promoters, livestock raising uses antibiotics to treat individual animals, as well as in routine dosing called "prevention and control" that protects whole herds. If antibiotics became useless, then animals would suffer: individual illnesses could not be treated, and if the crowded conditions in which most meat animals are raised were not changed, more diseases would spread.

But if the loss of antibiotics changes how livestock are raised, then farmers might be the ones to suffer. Other methods for protecting animals from disease—enlarging barns, cutting down on crowding, and delaying weaning so that immune systems have more time to develop—would be expensive to implement, and agriculture's profit margins are already thin. In 2002 economists for the National Pork Producers Council estimated that removing antibiotics from hog raising would force farmers to spend $4.50 more per pig, a cost that would be passed on to consumers.

H. Morgan Scott, a veterinary epidemiologist at Kansas State University, unpacked for me how antibiotics are used to control a major cattle illness, bovine respiratory disease. "If a rancher decides to wean their calves right off the cow in the fall and ship them, that's a risky process for the calf, and one of the things that permits that to continue is antibiotics," he said, adding, "If those antibiotics weren't available, either people would pay a much lower price for those same calves, or the rancher might retain them through the winter" while paying extra to feed them. That is, without antibiotics, those farmers would face either lower revenues or higher costs.

Livestock raising isn't the only aspect of food production that relies on antibiotics or that would be threatened if the drugs no longer worked. The drugs are routinely used in fish and shrimp farming, particularly in Asia, to protect against bacteria that spread in the pools where seafood is raised—and as a result, the aquaculture industry is struggling with antibiotic-resistant fish diseases and searching for alternatives. In the United States, antibiotics are used to control fruit diseases, but those protections are breaking down too. Last year, streptomycin-resistant fire blight, which in 2000 nearly destroyed Michigan's apple and pear industry, appeared for the first time in orchards in upstate New York, which is (after Michigan) one of the most important apple-growing states. "Our growers have never seen this, and they aren't prepared for it," says Herb Aldwinckle, a professor of plant pathology at Cornell University. "Our understanding is that there is one useful antibiotic left."

Is a post-antibiotic era inevitable? Possibly not—but not without change.

In countries such as Denmark, Norway, and the Netherlands, government regulation of medical and agricultural antibiotic use has helped curb bacteria's rapid evolution toward untreatability. But the United States has never been willing to institute such controls, and the free-market alternative of asking physicians and consumers to use antibiotics conservatively has been tried for decades without much success. As has the long effort to reduce farm antibiotic use: the FDA will soon issue new rules for agriculture, but they will be contained in a voluntary "guidance to industry," not a regulation with the force of law.

What might hold off the apocalypse for a while is more antibiotics—but first pharmaceutical companies will have to be lured back into a marketplace they have already deemed unrewarding. The need for new compounds could force the federal government to create drug-development incentives: patent extensions, for instance, or changes in the requirements for clinical trials. But whenever drug research revives, achieving a new compound takes at least ten years from concept to drugstore shelf. There will be no new drug to solve the problem soon—and given the relentlessness of bacterial evolution, none that can solve the problem forever. In the meantime, the medical industry is reviving the old-fashioned solution of rigorous hospital cleaning and also trying new ideas: building automatic scrutiny of prescriptions into computerized medical records and developing rapid tests to ensure that the drugs aren't prescribed when they are not needed. The threat of the end of antibiotics might even impel a reconsideration of phages, the individually brewed cocktails of viruses that were a mainstay of Soviet Union medical care during the Cold War. So far the FDA has allowed them into the U.S. market only as food-safety preparations, not as treatments for infections.

But for any of that to happen, the prospect of a post-antibiotic era has to be taken seriously, and those staring down the trend say that still seems unlikely. "Nobody relates to themselves lying in an ICU bed on a ventilator," says Rice of Brown University. "And after it happens, they generally want to forget it."

When I think of preventing this possible future, I reread my great-uncle's obit, weighing its old-fashioned language freighted with a small town's grief.

> The world is made up of "average" people, and that is probably why editorials are not written about any one of them. Yet among these average people, who are not "great" in political, social, religious, economic or other specialized fields, there are sometimes those who stand out above the rest: stand out for qualities that are intangible, that we can't put our finger on.
>
> Such a man was Joe McKenna, who died in the prime of life Friday. Joe was not one of the "greats." Yet few men, probably, have been mourned by more of their neighbors—mourned sincerely, and sorrowfully—than this red-haired young man.

I run my cursor over the image of the tattered newsprint, the frayed creases betraying the years that someone carried the clipping around. I picture my cousin's grandmother flattening the fragile scrap as gently as if she were stroking her brother's hot forehead, and reading the praise she must have known by heart, and folding it closed again. I remember the few stories I heard

from my father, of how Joe's death shattered his family, embitter-
ing my grandfather and turning their mother angry and cold.

I imagine what he might have thought—thirty years old, newly
married, adored by his siblings, thrilled for the excitement of his
job—if he had known that a few years later, his life could have
been saved in hours. I think he would have marveled at antibiotics,
and longed for them, and found our disrespect of them an enormous waste. As I do.

SETH MNOOKIN

The Return of Measles

FROM *The Boston Globe Magazine*

IF YOU WERE GOING to write down the most frightening infectious diseases you could think of, measles probably wouldn't be near the top of your list. Compared with the devastation of HIV/AIDS or the gruesome deaths caused by hemorrhagic fevers like Ebola, measles, with its four-day-long fevers and pervasive rashes, seems like nothing more than an annoyance.

But there is one thing that makes measles unique, and uniquely frightening to public health officials: it is the most infectious microbe in the world, with a transmission rate of around 90 percent. The fact that measles can live outside the human body for up to two hours makes a potential outbreak all the more menacing.

This explains the all-hands-on-deck response when officials with the Massachusetts Department of Public Health learned in late August that two unconnected patients—an infant who'd recently arrived in the United States and a foreign-born adult who'd recently traveled abroad—had visited area hospitals with active measles infections. Identifying the hundreds of people who'd potentially been exposed and then checking their vaccination status required, in the words of Dr. Larry Madoff, director of the state's Division of Epidemiology and Immunization, a "huge effort" on the part of dozens of state, local, and hospital employees.

Fortunately, there were no secondary infections this time around, a fact that is due in no small part to the impressive vaccine uptake rate in this state. It would be a mistake to assume this will always be the case: Massachusetts is seeing a surge in the number of unvaccinated children. Last year nearly 1,200 kids entered kindergarten with religious or philosophical vaccine exemptions, roughly double the total about a decade ago.

That mirrors what's happening across the country. What's so confounding is that many of the parents requesting exemptions for their children cite specious, disproven fears—such as that the vaccine could cause autism—many of which were based on a fraudulent, retracted study or fringe research published in non-peer-reviewed journals. And the rest of the country hasn't been as successful as Massachusetts in containing measles infections. Earlier this year, an intentionally unvaccinated seventeen-year-old from Brooklyn, New York, was infected with measles while on a trip to the United Kingdom. Because he lived in a community with a large number of other deliberately unvaccinated children, the virus quickly spread. By the time the outbreak was contained, fifty-eight people had been infected—making it the largest outbreak in the country in more than fifteen years. Nationwide, the Centers for Disease Control and Prevention reported 159 total cases between January and August, which puts 2013 on track to record the most domestic measles infections since the disease was declared eliminated from the United States in 2000.

In a country of more than 313 million, a couple hundred infections doesn't sound like a lot—and it's not. But you need only take a look across the Atlantic to find out how quickly measles can spread out of control. In 2007 there were just forty-four infections in France, a country where vaccination is recommended but not required. Over the next four years, more than twenty thousand additional cases were recorded. Nearly five thousand of these patients required hospitalization, and ten of them died.

As the containment efforts illustrate, the fact that there haven't been any recent deaths in the United States doesn't mean measles isn't having a real impact on the economy or on public health. One of the reasons Madoff oversaw an effort in Massachusetts to contact everyone who might have been exposed was to make sure they were OK. Another was to identify anyone who wasn't vaccinated so they could "isolate themselves and be out of work and out of school."

The state isn't releasing estimates for the total cost of these two infections, but a 2010 study in *Pediatrics* quantified the expense of containing a 2008 outbreak in San Diego in which 11 children were infected—and another 839 people were exposed. That cost the public sector $124,517, an average of more than $10,000 per infection. These are costs borne by all of us: every tax dollar spent containing measles is a dollar not spent on other public health initiatives.

Maybe you're not particularly civic-minded. If so, consider this: forty-eight children too young to be vaccinated in San Diego had to be quarantined, at an average cost of $775 per family. Another infant had to be hospitalized after being infected—a harrowing ordeal for that child's parents and one that rang up almost $15,000 in medical costs.

All of this is worth remembering the next time parents who don't vaccinate their children tell you they're making a purely personal choice. This is, of course, technically true, in the same sense that driving after having a few beers is a personal choice. As the mother of the ten-month-old hospitalized in San Diego said, if people want to make that choice, they should go live on an island with its own schools and doctors: "their own little infectious disease island."

JUSTIN NOBEL

Ants Go Marching

FROM *Nautilus*

FIVE WORLD WAR II bombers took off from a Florida airfield on October 5, 1967, to bomb the American South. An article that ran that morning in the Sarasota *Herald-Tribune* said that three B-17s and two PV-2s laden with 10,000 pounds of death-dealing cargo each would carry out their missions "with the City of Sarasota and eastern Manatee as their targets."

While the bombers were certainly at war, they weren't dropping explosives. Their enemy was a millimeters-long, brownish-red insect known to scientists as *Solenopsis invicta*, meaning "invincible ant," and known to lay people as fire ants, aka "ants from hell" and "them devils." The bombers were to unload mirex, a poison usually applied to grits, onto the critter.

By the late 1960s, the fire ant had been in the American South for more than thirty years. Southerners spoke of ruined crops, destroyed wildlife, and the ants' fiery sting. How much damage the ants had actually caused was uncertain, but it was enough for the U.S. Department of Agriculture (USDA) to declare war on the pest. During an eleven-year campaign, more than 143 million pounds of mirex were dropped across 77,220 square miles of land from Texas to Florida, costing close to $200 million.* The out-

---

* Mirex, also known as dechlorane, is a persistent organic pollutant (POP) that is now banned by the United Nations Environmental Programme and the U.S. Environmental Protection Agency. It is one of the original "dirty dozen" chemicals targeted for elimination by the international treaty signed at the Convention on Persistent Organic Pollutants in 2001.

come? The ants nearly doubled their range. The mirex, which was later found to be a carcinogen, persisted in the environment for decades, accumulating in birds' eggs, mammals' milk, and human tissues. The world's leading ant researcher, E. O. Wilson, dubbed the mirex program the "Vietnam of entomology."

Today, if you draw a line from Virginia Beach to Nashville to Abilene in west Texas, you'll find fire ants everywhere below it, as well as in Southern California.* The ants' annual impact on the economy, environment, and quality of life in the United States totals $6 billion, according to entomologists at Texas A&M University. In Texas alone they rack up $1.2 billion each year: $47 million at golf courses; $64 million at cemeteries (the ants love the open and slightly overgrown habitat around tombs); and as much as $255 million in the cattle industry. They cause other problems too. In Virginia Beach, a thirty-year-old former marine, Bradley Johnson, was stung by fire ants while working outside—and died of anaphylactic shock. On at least one occasion, fire ants invaded an elementary school in Tennessee to get candy stashed in kids' lockers. At Greystone Retirement Community in Huntsville, Alabama, a staffer found seventy-nine-year-old Lucille Devers covered in fire ants, which were crawling from her mouth, nose, ears, and hair: the ants frequently enter nursing homes, attracted by crumbs left in residents' beds.† And scientists anticipate that the ants will keep expanding their range. Climate change and crossbreeding with species more tolerant of cold may enable them to settle farther north.

I, on the contrary, moved south. Last August my girlfriend, Karen, and I sold most of our belongings, piled two cats, a terrier, and a pudgy brown Chihuahua named Jazzy-B into a minivan and drove from New York City to New Orleans. A few weeks later, we toasted our arrival with a picnic in City Park. As we rolled up our blanket, our legs suddenly caught fire. Under the glow of a street-

---

* There are five times more fire ants per acre in the United States than in their native South America. Fire ants cover 321 million acres in this country, across thirteen states and Puerto Rico, which adds up to 501,563 square miles. That's more than Germany, France, and the UK combined, or almost one-eighth of Europe.
† Lucille Devers, who was found swarmed by fire ants at the nursing home where she lived, was awarded $5.35 million by an Alabama jury. The $5.35 million award, returned on June 28, 2002, included $3.5 million in punitive damages, with Greystone and Terminix paying $1.75 million each.

light, we slipped off our pants to find our legs crawling with ants. I received more than two hundred stings, which formed welts the size of drink coasters, and my ears and throat swelled up. Half a dozen Benadryls later I was fine, except for hundreds of very itchy pustules. Luckily I wasn't anaphylactic, but between 0.5 and 5 percent of the U.S. population is.* For the most allergic, such stings can cause spasms of the bronchial muscles or coronary arteries, preventing oxygen from entering the bloodstream and causing death within minutes. When a few weeks later Jazzy-B stepped on an ant mound and howled, spending the rest of the day licking his paw, I was ready to declare my own war on the fire ant. But first I had to research my enemy.

The range of *Solenopsis invicta* covers a vast wetland in southern Brazil and Paraguay known as the Pantanal. Sometime in the early 1930s the ants stowed away in coffee sacks, soil, or hollow logs piled in the bottom of a cargo ship. The voyagers were probably just a handful of queens, each about the width of a thumbnail. They ate what they could find down in the hold—cockroaches, beetles, sugary cargo, and, when the pickings got slim, themselves, digesting their own wing muscles and fat reserves. Their ship may well have steamed past Rio de Janeiro, the mouth of the Amazon, and the lush peaks of the Antilles, all while their first batch of eggs gestated in their abdomens. In Mobile, Alabama, the ship docked, the sacks or soil or logs were unloaded, and the queens disembarked. Beneath the port's loading cranes and circling seagulls, perhaps in a patch of newly cut grass, the ants established their first colony: a mound of soil honeycombed with chambers and tunnels that ran as much as 4 feet deep. They like to build mounds in disturbed habitats such as the edge of a road, the side of a building, pastures, lawns, or near a busy port.† They eat pretty much anything— seeds, nectar, worms, weevils, butterflies, and even baby sea turtles, snakes, and alligators—catching the young as they hatch.

Colonies consist of queens, workers, and sexuals, also called

* Forty million people live in fire ant–infested areas, 30 to 60 percent of whom are stung annually by the ants, according to the USDA report "Integrated Management of Imported Fire Ants and Emerging Urban Pest Problems." It is estimated that 1 percent, 400,000 people, have an anaphylactic reaction.
† The ants are also drawn to electrical boxes; when one gets fried, a signal is released that brings others. The ants have been known to short out traffic lights and airport radar systems.

alates. Queens lay eggs that hatch into larvae—tiny, white, rice-like kernels that develop into adult ants. Alates are born at the end of winter or the beginning of spring and spend their lives fattening up and preparing for their mating flight. Workers begin life as nurses, grooming and feeding the brood and the queen. They move on to tasks like nest maintenance and sanitation. In their golden years, workers become foragers, handling the colony's most dangerous job, as it exposes them to predators and the elements. As the life in them winds down, workers act as their own pallbearers, lying down to die in a mass ant grave called the refuse pile.

To trace the ants' seemingly unstoppable march through the South, I decided to follow their path from where they first staked their claim to American soil. Just before I left, like an ominous sign from some myrmecological god, a mound swelled up in our backyard. I headed out on my odyssey nonetheless, undaunted.

Mobile, the twelfth largest port in the nation, which last year handled 26 million tons of cargo, greeted me with loading cranes that looked like characters from some gigantic industrial alphabet, a stench of diesel fuel, sea spray that hung in the air, and—sure enough—fire ant mounds. Near the port, cattails poked out the top of one. Its surface was lifeless, but when I kicked it lightly, seeking revenge for my itchy pustules, the creatures swarmed out. They might have been the great-great-great-great-great-great-great-great-great-great-great-grandchildren (queens live about seven years) of the group that had steamed here from South America back in the 1930s.

A short walk from the port was a neighborhood of arching oaks and trim ranch homes with well-clipped lawns. I stopped at 550 Charleston Street, the childhood home of E. O. Wilson—to ponder the fact that the planet's most famous ant biologist grew up at the epicenter of the planet's most famous ant invasion. In the weedy vacant lot next door, a young Wilson studied beetles, butterflies, spiders, and all manner of ants, including *Solenopsis invicta,* whose mound he first discovered in May 1942, at age twelve. "I still remember the species I found, in vivid detail," Wilson wrote in *Naturalist,* a book about his insect-happy childhood. Now the neighborhood appeared to be fire ant–less—thanks to the tireless, creative, and brutal efforts of its human inhabitants.

Sitting next to a table that held a rifle and a dead squirrel, Lonnie Rayford, a seventy-two-year-old farmer, told me he "done got bit" many a time. In his opinion, one that science does not really confirm, just plain grits were good enough to kill the pests. "Dig a hole and put grits in there, they gonna take it down to the bottom and give it to the queen," he said. "Them grits will kill them." Down the street, a husky man pulling weeds from his garden told me he used poison.

From Mobile I headed northeast to Montgomery, where I drove by the state capitol, a creamy white dome calling to mind a magnificent wedding cake. I stopped at Dixie Hardware to learn about fire ant control from the pros. A bearded worker named John showed me to aisle nine, which was an exterminator's nirvana.

"You got roach and ant killer, you got ant killer, you got Ant Max—it's some kind of trap," he said, holding up a rectangular red-and-yellow box that looked like it could have contained movie theater candies, except for a drawing of a vicious fire ant on the side. The best seller was a big orange bag that read "Spectracide Fire Ant Killer Mound Destroyer."

John's coworker Richard explained to me the challenges and intricacies of mass insecticide. The problem with poison was that ants often simply moved into the neighbor's yard, requiring neighborhood-wide poisoning efforts—in other words, it took a village to rid itself of the pests. Another trick involved two people shoveling ants from separate mounds into one another—according to Richard, the ants will assassinate the foreign queens. And then there is gasoline, apparently Alabama's preferred eradication method. "Take a broom handle, stick it down the mound, pour gas on it, and I know it's gonna burn whatever is in there," Richard said excitedly.

Eliminating fire ants seemed a bit like making cornbread: every Southerner had his own favorite recipe. By this time, my welts were long gone and I began to feel bad for the little ants. Especially since I understood that their inexorable spread was, in large part, our own fault.

Naturally, ants don't really spread that fast: during the 1930s and 1940s fire ants migrated out from Mobile at the rate of about 4 to 5 miles a year, a distance thought to be covered mostly in mating flights. On warm days following soaking spring or summer rains, alates exit the mound surrounded by a gang of worker bodyguards and launch into the air to mate—which they do in large clouds called mating swarms. Males inject females with a lifetime supply of sperm and then die. Newly mated queens land, kick off their wings, and scurry away to start new colonies, avoiding predators like dragonflies, which love to gobble their sperm-filled abdomens. Queens typically fly a few miles during mating flights, and if the wind is right, they can fly more than ten. But that's it.

Ants can also move when disturbed. In such a case an entire colony moves, but typically only a few feet, a migration initiated automatically by the workers. And a good many fire ant colonies across the South—known as polygyne colonies, meaning those with more than one queen—can actually move without mating flights, by budding off like yeast. A queen and some workers will simply wander off to start a new colony, but again, they won't go far. The most successful means for fire ant migration, however, is us. Humans.

In 1949 fire ants were mostly situated in a roughly 50-mile-wide radius around Mobile and in a few spots in central Alabama and Mississippi. But as Americans moved to the suburbs during the 1950s, desirous of white picket fences, front lawns, and ornamental shrubs, the nursery trade boomed. Soil, plants, and pots were shipped across states and counties. Hidden in the dirt caked on a bulldozer, in the root ball of a nursery plant, or in the bed of a pickup truck, a newly mated queen can travel hundreds of miles. Thanks to our horticultural aspirations, by 1957 *S. invicta* were in every southern state except Kentucky and the Virginias.

It was around then that the U.S. Department of Agriculture set out to eliminate the invaders.* In 1958 it established a quarantine that restricted the shipment of soil, nursery plants, and baled hay from areas infested with fire ants to those without, unless the products were first treated with insecticides. Between 1957 and 1962 the USDA coated 2.5 million acres with clay granules containing the insecticide heptachlor. It did indeed kill fire ants, but also

---

* Use of insecticide spray averages 4 fluid ounces per 1,000 square feet, which comes to 174,240 fluid ounces—or 1,361.25 gallons—of spray per acre. This equals 436,961,250,000 gallons for the entire affected region, or nearly 662,000 Olympic-size swimming pools full of insecticide spray each year in the United States.

blackbirds, quail, geese, frogs, dogs, cats, crabs, and millions of fish—a famous fiasco that helped inspire Rachel Carson's *Silent Spring*. Few at the USDA seemed to heed her warning, so the war continued. That entomological Vietnam, which lasted from 1964 to 1975, according to *The Fire Ants*, a book by myrmecologist Walter Tschinkel, left 24 to 33 percent of southerners with mirex in their body tissues.

Meanwhile, the invincible creatures nearly doubled their range, and some scientists speculated that the insecticide was the cause. The poison often eliminated all the ant species where it was applied, giving *S. invicta*, which is thought to be a better colony founder, a chance to become the dominant type upon recolonization. The secrets of their survival are not entirely known, but Tschinkel thinks it has something to do with large colonies, large numbers of alates, large dispersal distances, and a long mating flight season.

The fire ant front line runs through the middle of Tennessee. I continued north on the ants' path and pulled off the highway in McMinnville, the heart of the state's nursery industry and just inside the line. Here, nursery growers have to douse plants' roots, where newly mated queens sometimes cling undetected, with chlorpyrifos, a toxic and expensive insecticide. Tommy Boyd, co-owner of Boyd & Boyd Nursery, told me his workers did the task while wearing gloves and respirators. He refused to touch the stuff himself because the insecticide gave him horrible headaches. "I don't want to die over some chemical," Boyd told me.

I drove down a windy lane a short time later that brought me to Tennessee State University's Otis L. Floyd Nursery Research Center. Here I spoke with a gentle entomologist in a flannel shirt named Jason Oliver, who works on a less destructive and more natural ant eradication method: a project to introduce phorid flies, which prey on the ants back in Pantanal. In the late 1990s USDA researchers went to South America to capture different species of phorid flies, which they studied at an Animal and Plant Health Inspection Service (APHIS) lab in Gainesville, Florida, to determine the best one to introduce to various parts of the American South. Oliver's phorid flies are shipped from the lab to be released in Tennessee.

Phorids lay their eggs in the fire ant's thorax, and when the eggs hatch, the ant's head falls off.* Oliver showed me a video he had filmed in a petri dish, of a fly implanting eggs in an ant. A tiny dot buzzed about some ants; in the blink of an eye it brushed one of them, then continued flying. In that moment the fly had injected the eggs. It happened so quickly I asked Oliver to replay the video. It seemed like a perfect solution. But as impressive as the tiny fly was, it wasn't going to decapitate a whole colony, just make ants more scared to leave their nests. "We're always going to have fire ants," Oliver told me. "There is no way to eliminate them."

Weaving through rolling hills dotted with nurseries, I drove northwest from McMinnville, skirting the *S. invicta* front line: To the south, elementary schools, nursing homes, and greenhouses were being invaded, while to the north, the land was, supposedly, fire ant–free. The road took me through the area inundated in the catastrophic floods of May 2010. In Nashville, the Cumberland River rose 33 feet, flooding much of the city, including the Country Music Hall of Fame, the Grand Ole Opry House, and an untold number of fire ant mounds. Contrary to expectations, the ants didn't drown. Instead, they metastasized further.

The ants would have been well prepared, Louisiana State University entomologist Linda Hooper-Bui explained to me. In their homeland the Pantanal floods happen annually, so they learned to raft. As waters rise, ants evacuate lower tunnels and move higher in their mound, eventually gathering on top. Using hooks, called tarsi, on the tips of their legs, ants latch on to one another and create rafts. Late-stage larvae are covered in hooklike hairs that trap air, encasing them in bubbles. Worker ants stack these larvae three to five thick, forming pontoons that keep the rafts afloat. They place the queen in the middle with pupae and early-stage larvae, which don't have the crucial hairs to form the bubbles. Except for the clumps of eggs that workers carry in their jaws and a small amount of liquid food stored in their bodies that will last only a few days, the ants bring nothing aboard. Moreover, as the raft sets

---

* Several days after the phorid fly lays eggs in the thorax of a worker ant, the maggot releases a chemical that causes the ant to crumple over; it also loosens its head and front legs. The maggot then eats the contents of the ant's head and the head falls off. Other ants carry the body to the colony's refuse pile, including the head occupied by the maggot, which it uses as its pupal case. It emerges forty-five days later as an adult fly.

off, tipped into the water by the workers, they fling male alates overboard. If the raft is afloat longer than four days, the ants will begin to eat the brood—although not the ones used to make the raft. Rafts can hold together for as long as twenty-one days, surely long enough to survive the swollen Cumberland River.

Or at least this is what Steve Powell, an entomologist with the state of Tennessee, believes. In February of this year he received a call reporting fire ants in Cumberland City, a remote town about 80 miles north from Nashville and far above the front line. "There's no rhyme or reason why they should be there, so far from other fire ant infestations," Powell said. "If I had to guess, I'd say it was the flood."

Continuing my journey, I drove farther north into Cumberland City—a ramshackle collection of vine-ensnarled clapboard homes and shuttered storefronts on steep forested hills above the Cumberland River. On the edge of town sat a massive Tennessee Valley Authority coal plant with four 1,000-foot-tall smokestacks, some of the largest on earth. Beside the plant was a small restaurant with foggy windows, where farmers in overalls sat at low tables gobbling catfish and pork chops. I was looking for someone to speak with about the ants, expecting a thrilling validation of Powell's rafting theory. Instead I found an unsurprising surprise: Bailey Gafford, a weatherworn cattle farmer in mud-splattered boots, told me fire ants had been in Cumberland City since before the floods. He even suggested an innovative method of eradication I hadn't heard before: "Put snuff around the mound. They come out and you set 'em on fire."

On my way out of Cumberland City I stopped at a Civil War cemetery, where wildflowers cloaked weather-beaten tombs, and atop a small hill stood a dozen crude cabins, campground for a long-forgotten battle. Our own battle against the ant was still in full swing, however, and it didn't look like we were winning. The war on *S. invicta* suddenly fit into a broader picture, that of the Insecticide-Military-Industrial Complex. It was around the time that President Obama announced, concerning the nation's ongoing war on terrorism, that we cannot remain on "a perpetual wartime footing." The statement also seemed to apply to our war against the fire ants.

Besides, no matter how we perfect our tactics, these ultimate invaders seem to find new ways to advance. Somehow, *Solenopsis* *invicta* crossbred with *Solenopsis richteri,* another species that came from Pantanal by cargo ship to Mobile, in 1918.* Originally *S. invicta* drove out *S. richteri,* a less aggressive variety that prefers cooler weather and might have found the South too hot. But in the 1980s a hybrid of the two was discovered. No one knows exactly how the hybridization occurred—in certain parts of the Pantanal the two species' territories overlap, but they don't interbreed. Here they do, which is perhaps the most ominous sign. Purebred *S. invicta* don't survive more than three or four days in temperatures below freezing. Neither does *S. richteri.* But the hybrid survives freezing temperatures better than either pure species, according to a 2002 study in *Environmental Entomology.*

Could a hybrid—or perhaps a hybrid of a hybrid—one day push that front line farther north, to Washington, DC, and Philadelphia? What about New York City, with its 8 million residents as clueless as I once was? Recently the ants were brought by ship from the United States to Australia and also Taiwan; from there they invaded China. According to a 2004 paper in *Biological Invasions,* fire ants could potentially infest France, Italy, Greece, Japan, South Korea, Mexico, Central America, and large parts of Africa and India. Is there a way for us to coexist with the creatures? I was still hopeful.

Upon returning to New Orleans I discovered that the fire ant mound in our backyard had quadrupled in size. It now towered above the grass blades like Kilimanjaro. As I stood dumbfounded, contemplating what to do, a potato bug ascended the northern flank, an agile alpinist indeed, somehow not inducing a swarm. I wanted to be like the potato bug and coexist with the mound, dance my fingers upon its slopes without getting stung, but I knew it was impossible. Even if Karen and I somehow avoided the nest, there were our animals to think of—the cats, the terrier, and Jazzy-B, who at that moment was watching me from inside with curious dog eyes. Last time he got off easy, but if he received hundreds of stings, the poison would surely overwhelm his little Chihuahua system. The choice was clear. I would have to eradicate the mound.

The abundance of killing choices was what puzzled me: Pit rival colonies against each other, grits, gasoline, a broom and gasoline,

* More than 54,000 cargo ships are hustling goods around the world, and checking all of them for fire ants has proved impossible.

poison, and, if poison, which one? In the end I decided on a less ecologically invasive technique no one had mentioned except for Oliver—he muttered it under his breath toward the end of our conversation as if revealing a secret that embarrassed him: hot water. Ant Max and Spectracide Fire Ant Killer Mound Destroyer don't seem to want you to know about it.

With Jazzy-B and all the others safely inside, I emptied the cat litter box, dusted it with baby powder so the ants couldn't crawl up the sides, and shoveled the mound into it, quickly adding boiling water to fill it. A second pot of boiling water went into the open mound, to seep down into the interior tunnels and chambers, and then a third and a fourth, too, just to be sure I got the queen. There was an initial explosion of red, with numerous members of the brood being moved about, but the water cooked them instantly and, within the hour, the mound was dead.

I gathered my household armor and walked back toward the house, with far from a satisfying sense of achievement. I might have won this battle, but we were still losing the war. Inevitably, another mound would pop up. My "victory" was just a reprieve.

FRED PEARCE

TV as Birth Control

FROM *Conservation*

EARLIER THIS YEAR the Stanford human geographer Martin Lewis asked his students a simple question: How did they think U.S. family sizes compared with those in India? Between Indian and American women, who had the most children? It was, they replied, a no-brainer. Of course Indian women had more—they estimated twice as many. Lewis tried the question out on his academic colleagues. They thought much the same.

But it's not true. Indian women have more kids, it is true, but only marginally so: an average of 2.5 compared to 2.1. Within a generation, Indian women have halved the number of children they bear, and the numbers keep falling.

It's not that the population problem has gone away in India —yet. India has a lot of young women of childbearing age. Even if they have only two or three children each, that will still continue to push up the population, already over a billion, for a while yet. India will probably overtake China to become the world's most populous nation before 2030.

But India is defusing its population bomb. A fertility rate of 2.5 is only a smidgen above the long-term replacement level, which —allowing for girls who don't reach adulthood and some alarming rates of aborting female fetuses—is around 2.3. The end is in sight.

With most of the country still extremely poor, this is a triumph against all expectations. And it offers some intriguing clues to a question that has dogged demographers ever since Paul Ehrlich published his blockbuster book *The Population Bomb:* What can persuade poor people in developing countries to have fewer babies?

Taking time off from bemusing his students, Lewis decided to investigate. Being a geographer, he tackled the question with maps. He noted that within the overall rapid decline in Indian fertility, there continued to be great regional variations. So he mapped fertility in each Indian state and examined those patterns against the patterns for some of the demographers' favored drivers of lowered fertility. When he compared his maps, he found that variations in female education fit pretty well. So did economic wealth and the Human Development Index, which measures education, health, and income. The extent of urbanization looked like a pretty good match, too. But he also found that TV ownership tallied well with fertility across India. Not perfectly, he concluded, but as well as or better than the more standard indicators. A TV in the living room, in other words, might have the power to transform behavior in the bedroom.

Surprising? Maybe not. Lewis was following the lead of Robert Jensen and Emily Oster, development economists from the University of California, Los Angeles, and the University of Chicago, respectively. Four years ago, they reported compelling direct evidence from Indian villages that TV empowers women. They carried out detailed interviews in rural India as commercial cable and satellite TV were replacing the mostly dull and uninspired government programming.

The pair noted that the new diet of game shows, soap operas, and reality shows instantly became the villagers' main source of information about the outside world—especially about India's emerging urban ways of life. At the top of the ratings was *Kyunki Saas Bhi Kabhi Bahu Thi* (meaning "Because a mother-in-law was once also a daughter-in-law"). Based on life in the megacity of Mumbai, it was Asia's most watched TV show between 2000 and 2008 and was an eye opener for millions of rural Indian women. They saw their urban sisters working outside the home, running businesses, controlling money, and—crucially—achieving these things by having fewer children. Here was TV showing women a world of possibilities beyond bearing and raising children—a world in which small families are the key to a better life.

Soap operas give viewers time to develop strong emotional bonds with the characters, many of whom live as they do and experience the life traumas that they do. The impact of the new TV programming in rural India has been profound—and very positive, say Jensen and Oster. Their interviews revealed that when the new TV services arrived, women's autonomy increased while fertility and the acceptability of domestic violence toward women significantly decreased. Most of the changes occurred within a few months of the arrival of TV reception, when (as they put it) "interactions with the television are more intense." In fact, the researchers found that TV's influence on gender attitudes, social advancement, and fertility rates was equivalent to the impact of an extra five years of female education. This was the social revolution that delivered the geographical variation in Lewis's maps.

There is a history to using soap operas to cut fertility. It goes back to Mexico in the late 1970s, a time when the average Mexican woman had five or six babies and Mexico City was becoming the world's largest megacity. Miguel Sabido, then vice president of Televisa, the national TV network, developed a soap-opera format in which viewers were encouraged to relate to a character on the cusp of doing right or wrong—a "transitional character" whose ethical and practical dilemmas drove the plot lines.

His prime soap, or telenovela, *Acompáñame* ("Accompany Me") focused on the travails of a poor woman in a large family living in a rundown shack in a crime-ridden neighborhood. She wanted to break out and, after many travails and setbacks, did so by choosing contraception and limiting her family size. It was a morality tale, and nobody could mistake the message. The lessons were reinforced with an epilogue at the end of each episode, giving advice about family planning services.

Some accused Sabido of crude social engineering. But according to research by the country's National Family Planning Program, half a million women enrolled at family planning clinics while the soap was on, and contraceptive sales rose 23 percent in a year. A rash of similar soap operas with names such as *Vamos Juntos* ("We Go Together") and *Nosotros las Mujeres* ("We the Women") ran in Mexico throughout the 1980s. They were credited, at least anecdotally, with helping slash Mexican fertility rates. Thomas

Donnelly, USAID's local man at the time, concluded that they "have made the single most powerful contribution to the Mexican population success story."

The "Sabido Method" caught on. The American population campaigner William Ryerson later launched the Population Media Center in Shelburne, Vermont, to promote it worldwide. Among many copycat shows were Jamaica's *Naseberry Street*, which ran from 1985 through 1989, a period during which the fertility rate on the Caribbean island fell from 3.3 to 2.9; and Kenya's long-running *Tushauriane* ("Let's Talk About It"), launched in 1987. It topped the ratings and coincided with a cut in Kenyan fertility rates from 6.3 to 4.4 children.

But the precise contribution of such programs to falling fertility rates was always elusive. And more sophisticated TV viewers reacted against the crude propaganda of some Sabido soaps, with their clunky story lines and dialogue right out of government leaflets. The next wave of soap operas became more subtle: soft soap, if you will. Their narratives offered a realism that simply associated smaller families and use of family planning services with aspirational lifestyles, perfect family lives, and female emancipation. Thus it is not overt propaganda messages that really transform, so much as the window TV offers on a world previously unknown to most women. Seeing is believing.

A pioneering study into these more subliminal soap messages was published in 2008 by Eliana La Ferrara of Bocconi University in Milan and colleagues at the Inter-American Development Bank. She looked at more than a hundred Brazilian telenovelas broadcast from the 1970s to the 1990s, finding that 72 percent of the main female characters younger than fifty years old had no children, and a further 21 percent had only one child. At the start of this period, such a picture of Brazilian life was far from reality: the average woman had five children. But by the end, life had imitated art: Brazilian fertility today is 1.8 children per woman.

The core of La Ferrara's study looked at the impact of the slow geographical spread of programming by Rede Globo, a TV network that for a long time had a near monopoly on Brazilian telenovelas. She found that as the soaps reached each region and as the majority of the population tuned in, there was a discernible additional fall in fertility. She was able to demonstrate that the soaps were responsible for an average 6 percent fall in fertility overall and an 11 percent drop among women over thirty-five. What's more, fertility was falling across Brazil at a time when opposition from the Catholic Church kept government out of the business of family planning.

Of course, the messages about small families that are conveyed in soap operas and other popular media are in keeping with existing trends. They reflect and encourage emerging norms and aspirations about family size as much as they actually create them. But the evidence from India and Brazil—these two large countries that have been most successful in reducing fertility—is that at a certain stage in the process of change, the impact of soaps can be swift and dramatic.

We do not often hear about this reproductive revolution, says Lewis—even though many environmentalists argue that future population trends are absolutely critical to the future sustainability of humans on the planet. "It almost seems as though we have collectively decided to ignore this momentous transformation of human behavior."

Perhaps the disconnect comes from how we think about the problem. Understanding the current human population equation requires us to hold two opposing ideas in our heads at the same time. First, birthrates globally are falling almost everywhere. As John R. Wilmoth, the new head of the United Nations Department of Economic and Social Affairs Population Division, noted earlier this year: "Fertility levels have fallen substantially in most regions, far beyond what most observers expected fifty years ago." Half the world now lives in countries with fertility at or below the long-term replacement level of 2.3. That includes some devoutly Muslim countries, such as Iran, as well as many primarily rural countries where girls get a meager education and are often forced into marriage at a very young age. The average woman of the world today has 2.5 children, half what her mother and grandmother had.

The second fact is that despite falling fertility, there is a considerable increase in overall human numbers still to come. The UN's statisticians this year raised their estimate of the likely increase in the world's population in this century. They predict that rather than growing from today's 7.2 billion to 10.1 billion, the population will increase to almost 11 billion people by century's end. One reason for the apparent paradox of fast-falling fertility and continuing high population projections is that today's mothers were born at a time when women still had four or five children—most of whom, for the first time in history, grew to adulthood. As in India, this baby-boom generation is so numerous that, even with small families, they will inflate the numbers substantially before ceasing to be fertile. Also, to a lesser extent, rising life expectancy will increase the population.

But the *rate* of worldwide population increase is declining, from 2.1 percent in the 1960s to today's 1.2 percent. And demographers recently noted that we have reached what they call "peak child." The number of children younger than fifteen has stopped rising and is likely to start falling in the future.

Some countries, notably in southern and eastern Europe and in East Asia, have seen their fertility rates fall well below replacement level for a generation now. From Italy and Greece to Russia, Japan, and South Korea, this figure is at or below 1.5. Many argue that modern women in such countries are, in effect, on childbirth strike because their male partners, employers, and governments make it hard for them to combine parenthood with a career.

Will fertility rates in these countries recover to replacement levels? The UN currently believes they will, though no country has yet done so fully. Women may not do what the demographers think they will.

At the other end of the spectrum are questions about the fate of sub-Saharan Africa; here, despite some decline, fertility rates are often still above three children per woman. The rates reach seven children per woman in Niger and Mali in West Africa. Poor peasant farmers still often see children as a good investment in labor on the farm as well as an insurance policy for their old age.

If things go badly, Africa could end the century with as many as 3 billion people, compared to 1 billion today. The UN is currently of that view. It regards numerous African nations as being stuck in a poverty trap that is also a fertility trap. Recently it upped its estimates of both Africa's current fertility rates and its likely future rates. It reckons, for instance, that Africa's most populous nation, Nigeria, will surge from around 175 million today to more than 800 million before the century is out.

But others think that Africa in the twenty-first century could follow the Asian example from the late twentieth century, with fertility falling dramatically to replacement levels or below.

Who's right? Optimists point out that poverty no longer dooms countries to high fertility, as shown by examples such as Burma and Bangladesh. But pessimists say it is unlikely that hard-pressed African governments will be delivering either wealth or Asian-style education systems to their fast-growing citizenry anytime soon.

However, it may be a mistake to think that governments are key to reducing fertility. Evidence from both Asia and Latin America indicates that wider social forces are at work: women are making their own choices. Female emancipation is key.

This could happen in Africa, too. Time and again, surveys show that most African women also want smaller families. Even in the poorest countries, advances in health care and campaigns against major killer diseases mean that most kids grow to adulthood. Women do not need to produce large families in order to deliver the next generation. But they still must take charge of their lives sufficiently to achieve small families; this requires cultural change. And as we have seen, in much of the poor world these cultural drivers come not from books, schoolteachers, tribal chiefs, or even the words of priests and mullahs. As in the villages of India and Brazil, the most dynamic force in changing social mores in villages and slum communities is a box in the corner of the living room: the TV.

Television's spread throughout the world is extraordinary. There are today something like 1.4 billion TV sets worldwide, or roughly one for every five people. In Asia and Latin America, even the poorest are glued to the box. In India, states where the average daily income is below two dollars per person still have TV sets in more than half of households, say Jensen and Oster.

In Africa, however, the spread has been slower. Across the continent, a billion people have only around 50 million TV sets. Government television broadcasting is mostly amateurish and unattractive. Satellite dishes to access other stations remain rare outside southern Africa. Most people still listen primarily to the radio.

But growing evidence suggests that TV could be the catalyst for change. In Kenya, *Makutano Junction,* a TV soap funded by the

British government's Department for International Development
(DFID) together with the family planning agency Marie Stopes
International, is now in its twelfth series. The Sunday evening
prime-time staple has 7 million viewers in Kenya and many more
in neighboring Tanzania and Uganda. It addresses a range of social issues, with family planning at the forefront.

In a recent series, the female hero, Anna, loses her twins in
childbirth after failing to attend a clinic. She leaves the hospital
with her abject partner, Josiah, who declares, "You are right. I
should have come with you for checks at the clinic. Now we'll go
to family planning like you wanted."

Since *Makutano Junction* went on air in 2001, Kenya's fertility
rate has fallen from 5.0 to 3.8. DFID makes no great claim that *Makutano Junction* caused this drop. "The extent to which *MJ* is able to
contribute to actual changes in behaviour is difficult to establish,"
it says. But audience research shows that viewers appreciate and
act on "very specific and practical information" that *MJ* provides
—for instance, how to find Marie Stopes clinics.

Similarly, Ethiopia's Amharic-language radio show *Yeken Kignit*
("Looking Over One's Daily Life"), broadcast from 2002 to 2004,
coincided with an estimated fall in fertility rates in Amharic-speak-
ing areas from 5.4 to 4.3. That also coincided with a big increase
in demand for contraceptives in those areas, says the Population
Media Center's Ryerson.

We should not think the power of soaps is purely a developing-
world phenomenon. Many argue that soaps have played a role
in triggering changes in attitudes toward homosexuality and gay
marriage in Europe and North America, for instance. And even
Sabido-style programs are being tried in rich nations. Witness the
arrival of online soaps with overt messages, such as *East Los High* at
hulu.com. Launched in June 2013, the soap—funded by the Population Media Center with help from the California Family Health
Council—targets Latino teens with tales of a girl from a single-
parent household who struggles against temptation.

Looking back, it's ironic that many of the same activists warning
of the population bomb back in the 1960s were also telling people
to "kill your TV." They saw TV as a socially damaging technology,
bringing in its wake violence, destructive consumer desires, and

social dislocation. But TV can also be a force for good, giving isolated and underprivileged people—especially women—a window on different worlds and a sense that they can change their lives. It empowers and increases aspirations—and even delivers lower fertility rates. Could the humble soap save the world? Stay tuned.

COREY S. POWELL

The Madness of the Planets

FROM *Nautilus*

I AM A STAUNCH BELIEVER in leading with the bad news, so let me get straight to the point. Earth, our anchor and our solitary haven in a hostile universe, is in a precarious situation. The solar system around us is rife with instability.

Residents of Chelyabinsk, Russia, experienced this firsthand at 9:20 A.M. local time last February 15, when a 50-foot-wide asteroid slammed into Earth's atmosphere and exploded above the town, shattering windows, collapsing the roof of a local zinc factory, and sending more than a thousand people to the hospital with glass cuts and other injuries. Millions of people saw the videos of Comet ISON meeting a different but related cataclysmic fate as it took a swan dive past the sun on Thanksgiving Day. In the space of a few hours, the 4.5-billion-year-old comet was reduced to a cloud of sputtering rubble.

But these incidents are mere pixels in the sweeping picture emerging from the latest theories of how our solar system formed and evolved. Collisions and dislocations are not occasional anomalies; they are a fundamental cosmic condition.

"Things are not as simple as they were supposed to be, with the planets staying quiet forever," says Alessandro Morbidelli, a planetary dynamics expert at the Nice Observatory in France. "When the planets form, they don't know they have to form on good orbits to be stable for billions of years! So they are stable temporarily but are not stable for the lifetime of the star."

Translation: Earth was forged in chaos, lives in chaos, and may well end in chaos.

While Morbidelli is explaining all this to me in a cheery Italian accent, I cannot help fixating on the grim connotations of his last name. He and his scientific compatriots are amplifying a recent realization about our celestial home: instability is our natural state. For centuries, Isaac Newton and his followers envisioned a solar system that runs like divine clockwork. Only in the past decade have high-precision mathematical simulations shown just how wrong he was. Carl Sagan famously declared that "we're made of star stuff." Morbidelli has an equally profound message: we are made of cosmic chaos.

"You might take a trip around the galaxy, come back in 5 billion years, and say, hey, there is no Mercury anymore, and Earth is now on an eccentric orbit that is catastrophic for life," Morbidelli continues in his musical voice. It's easy to be cheery when talking about events that nobody alive will ever experience, I suppose, but his tone doesn't change a bit as he starts ticking off the parts of the solar system in flux.

"There are unstable populations in the asteroid belt, Kuiper Belt, and Oort Cloud," Morbidelli says. These instabilities are relics of the chaos in which Earth and the other planets formed. The Chelyabinsk meteor emerged from the asteroid belt, perhaps from a smashup that took place about 30,000 years ago. The Kuiper Belt, just beyond Neptune, and the Oort Cloud, a vast hive of dormant comets extending halfway out to the next star, can also send objects careening our way. That is where Comet ISON came from.

After the Chelyabinsk disaster, Morbidelli and a group of colleagues huddled to figure out the implications of all that instability. Meteor impacts seemed to be happening a lot more often than their models predicted. When they presented their updated results last month, they concluded that the true rate of Chelyabinsk-scale events is probably ten times as high as they had previously estimated. There is no clean separation between the unsettled past and the present.

Morbidelli's view replaces Newtonian clockwork with something more akin to quantum uncertainty, with everything defined by probabilities of survival. Over time, every object in the solar system that *can* be destroyed, scattered, or ejected *will* eventually be destroyed, scattered, or ejected. That is how Earth came to be. That is how it exists today.

*

The belief that Earth and the rest of the solar system were born in largely their present form—the arrangement and characteristics of the planets almost preordained—has deep, clinging roots in the history of science. It extends further than Newton, back to the influential writings of the thirteenth-century monk Johannes de Sacrobosco, who described the universe in terms of clean, geometric patterns. You could plausibly draw a line all the way to the perfect heavenly spheres of Aristotle. Amazingly, the same basic philosophy (built atop a very different scientific foundation) persisted well into the space age.

In retrospect, that simple and comforting view had begun to unravel long before most scientists recognized what was happening. As far back as the 1970s, theoretical models trying to simulate the formation of the solar system kept coming up with an unwanted result: the planets migrated wildly, toward and away from the sun, making a royal mess of things in the process. "It would come out of pen-and-paper theory work, but it was immediately ignored," says Kevin Walsh, a leading researcher in solar system dynamics at Southwest Research Institute in Boulder, Colorado. "It showed up as throwaway comments in some papers, but no one seriously proposed that planetary migration was an important process."

The idea languished for a couple decades until a series of scene-shifting discoveries forced the theorists to reconsider. In 1995 two Swiss astronomers detected 51 Pegasi b, a planet orbiting a dim yellow star located 50 light-years away in the constellation Pegasus. It was the first world found orbiting another star similar to the sun. The planet is a gassy, Jupiter-size giant, the kind of world that theoretically can form only in the cool regions far from its parent star. But there was 51 Pegasi b, hugging close in a searing-hot orbit. The only sensible explanation astronomers could come up with was that the planet had formed far out and then somehow shifted sharply inward. They dubbed this puzzling world a "hot Jupiter" and waited to see if it was a fluke.

It was not. Within a year, a competing American team discovered two more hot Jupiters; several dozen similar worlds are now catalogued. As astronomers got better at searching, they started to find a number of other improbable planets. Some were in highly oval orbits; some revolved around their stars at a steep angle, or even backward. Such arrangements did not seem physically possible unless the planets had migrated dramatically at some point. If that process happened around other stars, it could have happened here as well. "And that's when it really started to kick off," Walsh says.

In 2009 Walsh was a postdoctoral fellow at the Nice Observatory, where he collaborated with Morbidelli. Walsh was already an expert in solar system dynamics; now he became fascinated by the concept of migrating planets and buried himself in the details of how the process would work. He focused on the earliest stages of solar system formation while Morbidelli tackled a later, secondary instability.

Getting planets to move is extremely easy in mathematical models of a newborn solar system. The challenge—as those pen-and-paper theorists of the 1970s had discovered—was finding ways for planets *not* to move. Data from the Hubble Space Telescope and other great observatories show that in the big picture infant planets emerge from a swirling disk of gas and dust around a just-formed star, known as a protoplanetary nebula. For the first few million years, planets are little more than debris bobbing on the waves in the disk.

"That nebula outweighs the planets about a thousand to one, so the gas can push the planets around really dramatically," Walsh says. As a result, he realized, the early solar system must have been more like bumper cars than clockwork. He also saw that if he fully embraced the idea of instability and took it to its logical conclusions, he could account for many aspects of the solar system that had previously defied easy explanation: Why is Mars so small? How did the asteroid belt form? And above all, why is Earth's chemical makeup so different from what was predicted by the original formation models?

Walsh knit his ideas into a theory he calls the Grand Tack, which creates a startlingly new narrative of how Earth and the other planets formed. At present, Jupiter's orbit is 5.2 times wider than Earth's. It is also sticking to its 11.8-year orbit like a metronome.

But according to Walsh, Jupiter actually formed quite a bit farther out and then, during the solar system's initial 5 million years, executed a series of dramatic swoops. First it spiraled inward to the place where Mars is now (about 1.5 times the Earth-sun distance), as the dense gas in the nebula dragged it toward the sun. Then it migrated out past its current location, yanked by the gravitational influence of the newly formed planet Saturn. The whole process took about 500,000 years—an eternity in human terms, but blazingly fast for the solar system, which is 4.6 billion years old.

So what happens, I ask, when a planet that size goes on the prowl? "Oh, it raises hell!" Walsh replies. "That's a really big planet and it's moving all over the place. It acts like a giant snowplow and essentially wipes out everything in its way."

Fortunately for us, Earth had not yet formed when Jupiter was on the move; if it had, our planet might have plunged into the sun or spun off into dark oblivion. The giant planet's influence on the inner solar system, where we live, was more indirect. Most of the action happened on the outbound track, when Jupiter rammed through thick swarms of icy comets and asteroids. That snowplow effect sent those water-rich objects raining down on Earth just as it was beginning to grow. "The bulk of the water that we see on Earth is a result of the scattering from Jupiter's outward migration," Walsh says. Whenever you take a swim, or just take a drink, you are benefitting from the solar system's foundational instability.

The migration of Jupiter reshaped the solar system in many other ways. It cleared out the original asteroid belt on the way in and filled it with new objects on the way out. It reorganized the distribution of comets. It stunted the growth of Mars, making it into the cold and nearly airless world it is today—bad luck for prospective Martians. At the same time, Jupiter deposited enough material closer to Earth that our planet ended up colliding with one of the leftover planetary cores. The moon is thought to have formed from the wreckage of that cataclysm.

After Jupiter finished its wanderings, the solar system looked stable, but the appearance was superficial. Instead, it had set the stage for a second great upheaval, one that has been baffling scientists for half a century.

NASA's Apollo missions achieved many notable things, but for the purposes of this story their greatest legacy was bringing back 842 pounds of moon rocks. On Earth, almost all evidence of the solar system's unruly early history has been worn away by erosion, biological activity, and the slow dancing of the continents. On the moon, there is no erasure. The moon never forgets.

During the 1970s, even as NASA was retreating from the majestic Apollo missions to a rickety Skylab, several groups of researchers set out to decode the lunar samples brought back by the astronauts. The moon's surface contains an intact chemical record of all the asteroids that have pummeled it over the eons. Planetary scientists expected to find a steady progression from disorder to order: lots of impacts right after the formation of the solar system, then a rapid tapering off as the moon (like Earth and the other planets) mopped up the last bits of debris. That is not at all the story written in the rocks.

When three geochemists—Fouad Tera, Dimitri Papanastassiou, and Gerald Wasserburg—sifted thoroughly through the lunar material, they saw instead that most of the material created by impacts had an age of about 3.9 billion years. The moon apparently experienced another intense barrage of asteroids at that time, a full 700 million years after the formation of the solar system. The researchers called it the "terminal lunar cataclysm"; now it's known as the Late Heavy Bombardment. Either way, it stayed on the books for decades as one of the solar system's biggest mysteries.

Around 2005, Morbidelli decided to take a crack at solving it. In conjunction with three other researchers, including Harold Levison, Walsh's neighbor and collaborator at the Southwest Research Institute, he wrote up a series of papers in which he connected the Late Heavy Bombardment to a previously unknown, belated instability in the formation of the solar system. The resulting "Nice model" (that's Nice as in the French town where Morbidelli works) is now the most widely accepted explanation for the solar system's second wave of devastating impacts.

According to this model, the solar system never quite found its steady groove after Jupiter migrated back out and the sun blew out its birth nebula. An enormous cloud of comets orbiting the sun beyond Neptune—the region that now marks the Kuiper Belt—was slowly but inexorably working gravitational mischief.

At first Neptune's orbit was synchronized with Jupiter's, a pattern called a resonance. Jupiter circled the sun three times, perhaps, for each one orbit of Neptune. Resonances tend to keep things stable. Over hundreds of millions of years, however, that cometary cloud dragged Neptune into a new orbit. "When Neptune gets off resonance with Jupiter, the system goes 'boom,' completely unstable. Then the violent evolution starts," Morbidelli says. Neptune migrated out, flinging comets inward; those comets reached Jupiter, which batted them even farther out; Jupiter, in response, migrated inward.

In the end, Saturn and Uranus, as well as Neptune, moved into more distant orbits. Jupiter settled into its present, closer one. In one version of the theory, developed by Morbidelli's colleague David Nesvorny at the Southwest Research Institute, our solar system originally had a fifth giant planet that got ejected entirely during this commotion; if so, it is currently wandering alone among the stars. Most comets got exiled to the Oort Cloud, far beyond the planets. Many other comets and asteroids went careening closer to the sun, where great numbers of them smacked into the moon, Earth, and the other inner planets.

Although almost all traces of this hellish era have vanished from Earth, some fleeting bits remain. Bruce Simonson of Oberlin College is tracking down extremely ancient impacts from the most distinctive evidence they leave behind: glass spherules the size of BBs (created from rock melted by the asteroid or comet) and elevated concentrations of the element iridium (much more common in meteorites than in Earth's surface). Earth's two oldest rock beds, one in western Australia and the other in South Africa, preserve records going back at least 3.4 billion years. For the past two decades, Simonson has been prospecting there for signs of what the Late Heavy Bombardment did to our planet.

One of his most intriguing results is that the rain of asteroids may have continued for a staggeringly long time, until 2.5 billion years ago or even more recently. "There's evidence it was a more gradual ramp-down, and we think it's convincing evidence, but of course it's ours," Simonson says. If he's right, regular asteroid blows were occurring well into the era of life on Earth, which began around 3.5 billion years ago. He also sees signs of what he technically calls "big-ass impacts," many times bigger than the impact that helped kill off the dinosaurs.

Surprisingly, Simonson thinks life soldiered on just fine amid all the falling rocks. "I'm not a big fan of impacts and extinction," he confides. "The only one where we have clear coincidence between impact and a change is the end-Cretaceous [when the dinosaurs went extinct]." He thinks that, overall, extinctions are more likely to have been caused by enormous volcanic eruptions and by the changing configuration of the continents and oceans. That's a little comfort, at least, considering that everywhere on Earth where he looks, he finds hints of ancient asteroid blasts.

At the time of the Late Heavy Bombardment, asteroids were hitting Earth at least a thousand times as often as they do now. Could something like that happen again? No, both Morbidelli and Walsh answer without hesitation. The first two planetary rearrangements cleared out 99.9 percent of the asteroid belt and Kuiper Belt; there is just not enough left to re-create the kind of chaos that ruled 3.9 billion years ago.

So are we home free? No again. "The terrestrial planets, they are not totally stable," Morbidelli says. That instantly captures my attention: Earth is one of the four terrestrial planets. "Mercury is on the edge of the instability, and it could go nuts, start to encounter Venus, then the orbits of Venus and the Earth could become unstable themselves." From there, Venus could collide with Earth, or Earth could go careening off on a totally new orbit, sterilizing the planet. The odds are not great, but they're not all that small either—about 1 percent over the next few billion years.

I question Morbidelli to make sure I'm understating him correctly. A 1 percent chance of disaster is surprisingly high odds in the cosmic-doomsday business. He sets down the phone for a moment and I hear him in the distance, double-checking with someone else in his office ("Do you know the probability that Mercury gets crazy?"). Then he's back on the line: "Yes, 1 percent." And he warns that the subtle divergences that would set the whole cataclysm in motion are like the weather, chaotic and impossible to forecast far in advance. They could be building up right now.

We are back to a probabilistic view of the solar system, in which nature builds some inherent uncertainty into the system. "It may be that instabilities are just a natural part of life for planetary systems as complicated as ours, and chaos keeps us from really understanding it," Walsh says.

There are definitely other, less catastrophic, and more comprehensible forms of instability at work right now. Comets still leak out of the region around the Kuiper Belt, a lingering hangover from the Late Heavy Bombardment. Comets in the Oort Cloud get sent flying by passing stars. In addition to gravitational mischief by the planets, a slight pressure created by sunlight, known as the

Yarkovsky Effect, keeps shifting the paths of the asteroids, guaranteeing that the risk of impact will never go away. A year ago, the standard line was that events like Chelyabinsk happen once every 200 or 300 years. The updated estimate is a couple times a century, maybe more often.

But Morbidelli is not at all gloomy or apocalyptic about his work. The more I speak to him, the more I absorb his perspective. Instability is a mechanism that transforms things from generic and boring into particular and interesting—whether those things are people or planetary systems. "If you want to describe the global evolution of a person, well, he's born and then he dies. That's it," Morbidelli says. "If you want to describe a specific person in detail, I cannot do it in a general scheme. There is a general scheme, but there are plenty of specific ramifications that drive you to be the person that you are. For a planetary system it is the same thing. That's chaos: extreme sensitivity to tiny changes."

To Morbidelli, we're not at war with a hostile universe, we're part of it. The astronomer has clearly spent a lot of time contemplating the personal implications of his work. "Planetary systems evolve by instable steps until they find their final peace; it is almost a Buddhist view of the universe," he concludes, as lyrical as ever. "Everything evolves to wisdom and peace—and stability—through big revolutionary events."

ROY SCRANTON

Learning How to Die in the Anthropocene

FROM *The New York Times*

I.

Driving into Iraq just after the 2003 invasion felt like driving into the future. We convoyed all day, all night, past army checkpoints and burned-out tanks, till in the blue dawn Baghdad rose from the desert like a vision of hell: flames licked the bruised sky from the tops of refinery towers, cyclopean monuments bulged and leaned against the horizon, broken overpasses swooped and fell over ruined suburbs, bombed factories, and narrow ancient streets.

With "shock and awe," our military had unleashed the end of the world on a city of 6 million—a city about the same size as Houston or Washington. The infrastructure was totaled: water, power, traffic, markets, and security fell to anarchy and local rule. The city's secular middle class was disappearing, squeezed out between gangsters, profiteers, fundamentalists, and soldiers. The government was going down, walls were going up, tribal lines were being drawn, and brutal hierarchies savagely established.

I was a private in the United States Army. This strange, precarious world was my new home. If I survived.

Two and a half years later, safe and lazy back in Fort Sill, Oklahoma, I thought I had made it out. Then I watched on television as Hurricane Katrina hit New Orleans. This time it was the weather that brought shock and awe, but I saw the same chaos and urban collapse I'd seen in Baghdad, the same failure of planning and the same tide of anarchy. The 82nd Airborne hit the ground, took over strategic points, and patrolled streets now under de facto martial law. My unit was put on alert to prepare for riot control operations. The grim future I'd seen in Baghdad was coming home: not terrorism, not even WMD's, but a civilization in collapse, with a crippled infrastructure, unable to recuperate from shocks to its system.

And today, with recovery still going on more than a year after Sandy and many critics arguing that the Eastern Seaboard is no more prepared for a huge weather event than we were last November, it's clear that that future's not going away.

This March, Admiral Samuel J. Locklear III, the commander of the United States Pacific Command, told security and foreign policy specialists in Cambridge, Massachusetts, that global climate change was the greatest threat the United States faced—more dangerous than terrorism, Chinese hackers, and North Korean nuclear missiles. Upheaval from increased temperatures, rising seas, and radical destabilization "is probably the most likely thing that is going to happen . . . ," he said, "that will cripple the security environment, probably more likely than the other scenarios we all often talk about."

Locklear's not alone. Tom Donilon, the national security adviser, said much the same thing in April, speaking to an audience at Columbia's new Center on Global Energy Policy. James Clapper, the director of national intelligence, told the Senate in March that "extreme weather events (floods, droughts, heat waves) will increasingly disrupt food and energy markets, exacerbating state weakness, forcing human migrations, and triggering riots, civil disobedience, and vandalism."

On the civilian side, the World Bank's recent report, *Turn Down the Heat: Climate Extremes, Regional Impacts, and the Case for Resilience,* offers a dire prognosis for the effects of global warming, which climatologists now predict will raise global temperatures by 3.6 degrees Fahrenheit within a generation and 7.2 degrees Fahrenheit within ninety years. Projections from researchers at the University of Hawaii find us dealing with "historically unprecedented" climates as soon as 2047. The climate scientist James Hansen, formerly with NASA, has argued that we face an "apocalyptic" future. This grim view is seconded by researchers worldwide, including

Anders Levermann, Paul and Anne Ehrlich, Lonnie Thompson, and many, many, many others.

This chorus of Jeremiahs predicts a radically transformed global climate forcing widespread upheaval—not possibly, not potentially, but inevitably. We have passed the point of no return. From the point of view of policy experts, climate scientists, and national security officials, the question is no longer whether global warming exists or how we might stop it, but how we are going to deal with it.

II.

There's a word for this new era we live in: the Anthropocene. This term, taken up by geologists, pondered by intellectuals, and discussed in the pages of publications such as *The Economist* and the *New York Times,* represents the idea that we have entered a new epoch in the earth's geological history, one characterized by the arrival of the human species as a geological force. The biologist Eugene F. Stoermer and the Nobel-Prize–winning chemist Paul Crutzen advanced the term in 2000, and it has steadily gained acceptance as evidence has increasingly mounted that the changes wrought by global warming will affect not just the world's climate and biological diversity, but its very geology—and not just for a few centuries but for millennia. The geophysicist David Archer's 2009 book, *The Long Thaw: How Humans Are Changing the Next 100,000 Years of Earth's Climate,* lays out a clear and concise argument for how huge concentrations of carbon dioxide in the atmosphere and melting ice will radically transform the planet, beyond freak storms and warmer summers, beyond any foreseeable future.

The Stratigraphy Commission of the Geological Society of London—the scientists responsible for pinning the "golden spikes" that demarcate geological epochs such as the Pliocene, Pleistocene, and Holocene—has adopted Anthropocene as a term deserving further consideration, "significant on the scale of Earth history." Working groups are discussing what level of geological timescale it might be (an "epoch" like the Holocene, or merely an "age" like the Calabrian), and at what date we might say it began. The beginning of the Great Acceleration, in the middle of the twentieth century? The beginning of the Industrial Revolution, around 1800? The advent of agriculture?

The challenge the Anthropocene poses is a challenge not just to national security, to food and energy markets, or to our "way of life"—though these challenges are all real, profound, and inescapable. The greatest challenge the Anthropocene poses may be to our sense of what it means to be human. Within one hundred years—within three to five generations—we will face average temperatures 7 degrees Fahrenheit higher than today, rising seas at least 3 to 10 feet higher, and worldwide shifts in crop belts, growing seasons, and population centers. Within a thousand years, unless we stop emitting greenhouse gases wholesale right now, humans will be living in a climate the earth hasn't seen since the Pliocene, 3 million years ago, when oceans were 75 feet higher than they are today. We face the imminent collapse of the agricultural, shipping, and energy networks upon which the global economy depends, a large-scale die-off in the biosphere that's already well on its way, and our own possible extinction. If *Homo sapiens* (or some genetically modified variant) survives the next millennium, it will be survival in a world unrecognizably different from the one we have inhabited.

Geological time scales, civilizational collapse, and species extinction give rise to profound problems that humanities scholars and academic philosophers, with their taste for fine-grained analysis, esoteric debates, and archival marginalia, might seem remarkably ill suited to address. After all, how will thinking about Kant help us trap carbon dioxide? Can arguments between object-oriented ontology and historical materialism protect honeybees from colony collapse disorder? Are ancient Greek philosophers, medieval theologians, and contemporary metaphysicians going to keep Bangladesh from being inundated by rising oceans?

Of course not. But the biggest problems the Anthropocene poses are precisely those that have always been at the root of humanistic and philosophical questioning: "What does it mean to be human?" and "What does it mean to live?" In the epoch of the Anthropocene, the question of individual mortality—"What does my life mean in the face of death?"—is universalized and framed in scales that boggle the imagination. What does human existence mean against 100,000 years of climate change? What does one life mean in the face of species death or the collapse of global civilization? How do we make meaningful choices in the shadow of our inevitable end?

These questions have no logical or empirical answers. They are philosophical problems par excellence. Many thinkers, including Cicero, Montaigne, Karl Jaspers, and "The Stone's" own Simon Critchley, have argued that studying philosophy is learning how to die. If that's true, then we have entered humanity's most philosophical age—for this is precisely the problem of the Anthropocene. The rub is that now we have to learn how to die not as individuals but as a civilization.

III.

Learning how to die isn't easy. In Iraq, at the beginning, I was terrified by the idea. Baghdad seemed incredibly dangerous, even though statistically I was pretty safe. We got shot at and mortared, and IEDs laced every highway, but I had good armor, we had a great medic, and we were part of the most powerful military the world had ever seen. The odds were good I would come home. Maybe wounded, but probably alive. Every day I went out on mission, though, I looked down the barrel of the future and saw a dark, empty hole.

"For the soldier death is the future, the future his profession assigns him," wrote Simone Weil in her remarkable meditation on war, *The Iliad, or the Poem of Force.* "Yet the idea of man's having death for a future is abhorrent to nature. Once the experience of war makes visible the possibility of death that lies locked up in each moment, our thoughts cannot travel from one day to the next without meeting death's face." That was the face I saw in the mirror, and its gaze nearly paralyzed me.

I found my way forward through an eighteenth-century Samurai manual, Yamamoto Tsunetomo's *Hagakure,* which commanded: "Meditation on inevitable death should be performed daily." Instead of fearing my end, I owned it. Every morning, after doing maintenance on my Humvee, I'd imagine getting blown up by an IED, shot by a sniper, burned to death, run over by a tank, torn apart by dogs, captured and beheaded, and succumbing to dysentery. Then, before we rolled out through the gate, I'd tell myself that I didn't need to worry, because I was already dead. The only thing that mattered was that I did my best to make sure everyone else came back alive. "If by setting one's heart right every morning and evening, one is able to live as though his body were already dead," wrote Tsunetomo, "he gains freedom in the Way."

I got through my tour in Iraq one day at a time, meditating each morning on my inevitable end. When I left Iraq and came back stateside, I thought I'd left that future behind. Then I saw it come home in the chaos that was unleashed after Katrina hit New Orleans. And then I saw it again when Sandy battered New York and New Jersey: government agencies failed to move quickly enough, and volunteer groups like Team Rubicon had to step in to manage disaster relief.

Now, when I look into our future—into the Anthropocene—I see water rising up to wash out lower Manhattan. I see food riots, hurricanes, and climate refugees. I see 82nd Airborne soldiers shooting looters. I see grid failure, wrecked harbors, Fukushima waste, and plagues. I see Baghdad. I see the Rockaways. I see a strange, precarious world.

Our new home.

The human psyche naturally rebels against the idea of its end. Likewise, civilizations have throughout history marched blindly toward disaster, because humans are wired to believe that tomorrow will be much like today—it is unnatural for us to think that this way of life, this present moment, this order of things, is not stable and permanent. Across the world today, our actions testify to our belief that we can go on like this forever, burning oil, poisoning the seas, killing off other species, pumping carbon into the air, ignoring the ominous silence of our coal mine canaries in favor of the unending robotic tweets of our new digital imaginarium. Yet the reality of global climate change is going to keep intruding on our fantasies of perpetual growth, permanent innovation, and endless energy, just as the reality of mortality shocks our casual faith in permanence.

The biggest problem climate change poses isn't how the Department of Defense should plan for resource wars, or how we should put up seawalls to protect Alphabet City, or when we should evacuate Hoboken. It won't be addressed by buying a Prius, signing a treaty, or turning off the air conditioning. The biggest problem we face is a philosophical one: understanding that this civilization is already dead. The sooner we confront this problem, and the sooner we realize there's nothing we can do to save ourselves, the sooner we can get down to the hard work of adapting, with mortal humility, to our new reality.

The choice is a clear one. We can continue acting as if tomorrow will be just like yesterday, growing less and less prepared for each new disaster as it comes, and more and more desperately invested in a life we can't sustain. Or we can learn to see each day as the death of what came before, freeing ourselves to deal with whatever problems the present offers without attachment or fear.

If we want to learn to live in the Anthropocene, we must first learn how to die.

KATE SHEPPARD

Under Water

FROM *Mother Jones*

Two months after Hurricane Sandy pummeled New York City,
Battery Park is again humming with tourists and hustlers, guys selling foam Statue of Liberty crowns, and commuters shuffling off
the Staten Island ferry. On a winter day when the bright sun takes
the edge off a frigid harbor breeze, it's hard to imagine all this
under water. But if you look closely, there are hints that not everything is back to normal.

Take the boarded-up entrance to the new South Ferry subway
station at the end of the No. 1 line. The metal structure covering the stairwell is dotted with rust and streaked with salt, tracing
the high-water mark at 13.88 feet above the low-tide line—a level
that surpassed all historical floods by nearly 4 feet. The saltwater
submerged the station, turning it into a "large fish tank," as the
former Metropolitan Transportation Authority chairman Joseph
Lhota put it, corroding the signals and ruining the interior. While
the city reopened the old station in early April, the newer one is
expected to remain closed to the public for as long as three years.

Before the storm, South Ferry was easily one of the more extravagant stations in the city, refurbished to the tune of $545 mil-
lion in 2009 and praised by the former MTA CEO Elliot Sander as
"artistically beautiful and highly functional." Just three years later,
the city is poised to spend more than that amount fixing it. Some
have argued that South Ferry shouldn't be reopened at all.

The destruction in Battery Park could be seen as simple misfor-
tune: after all, city planners couldn't have known that within a few

years the beautiful new station would be submerged in the most destructive storm to ever hit New York City.

Except for one thing: they sort of *did* know. Back in February 2009, a month before the station was unveiled, a major report from the New York City Panel on Climate Change—which Mayor Michael Bloomberg convened to inform the city's climate adaptation planning—warned that global warming and sea-level rise were increasing the likelihood that New York City would be paralyzed by major flooding. "Of course it flooded," said George Deodatis, a civil engineer at Columbia University. "They spent a lot of money, but they didn't put in any floodgates or any protection."

And it wasn't just one warning. Eight years before the Panel on Climate Change's report, an assessment of global warming's impacts in New York City had also cautioned of potential flooding. "Basically pretty much everything that we projected happened," says Cynthia Rosenzweig, a senior research scientist at NASA's Goddard Institute for Space Studies, the cochair of the Panel on Climate Change and coauthor of that 2001 report.

Scientists often refer to the "hundred-year flood," the highest water level expected over the course of a century. But with sea levels rising along the East Coast—a natural phenomenon accelerated by climate change—scientists project that in our lifetimes what was once considered a hundred-year flood will happen every three to twenty years. And truly catastrophic storms will do damage unimaginable today. "With the exact same Sandy one hundred years from now," Deodatis says, "if you have, say, five feet of sea-level rise, it's going to be much more devastating."

Roughly 123 million of us—39 percent of the U.S. population—dwell in coastal counties. And that spells trouble: 50 percent of the nation's shorelines, 11,200 miles in all, are highly vulnerable to sea-level rise, according to the National Oceanic and Atmospheric Administration. And the problem isn't so much that the surf laps a few inches higher: it's what happens to all that extra water during a storm.

We're already getting a taste of what this will mean. Hurricane Sandy is expected to cost the federal government $60 billion. Over the past three years, ten other storms have each caused more than $1 billion in damage. In 2011 the federal government declared a record ninety-four weather-related major disasters, from hurri-

canes to wildfires.* The United States averaged fifty-six such disasters per year from 2000 to 2010, and a mere eighteen a year in the
1960s.

The consequences for the federal budget are staggering. In just
the past two years, natural disasters have cost the Treasury $188
billion—nearly $2 billion a week. The National Flood Insurance
Program (NFIP), which covers more than $1 trillion in assets, is
one of the nation's largest fiscal liabilities. The program went $16
billion in the hole on Hurricane Katrina, and after Sandy it will be
at least $25 billion in debt—a deficit unlikely ever to be fixed.

Meanwhile, Washington is stuck in an endless cycle of disaster
response. The U.S. government spends billions of dollars on disasters after they happen, but it pinches pennies when it comes
to preparing for them. And both federal and state policies create incentives for people to build and rebuild in increasingly risky
coastal areas. "The large fiscal machinery at the federal level is
cranking ahead as if there's no sea-level rise, and as if Sandy never
happened," says David Conrad, a water consultant who has been
working on flood policy for decades. "This issue is moving so much
faster than the governmental apparatus right now."

Put another way, we're already deep under water.

Some 350 miles south of the South Ferry subway station, on Virginia's Middle Peninsula, is Gloucester County. The red-brick homes
and matching sidewalks give the county seat, Gloucester Court-
house, a colonial feel. The county was the site of Werowocomoco,
the capital of the Powhatan Confederacy, and just across the York
River to the southwest are Jamestown, Yorktown, and Williamsburg, some of the earliest English settlements in North America.
Europeans first settled the county in 1651, and some local families
date back just about that far. In the more remote necks on the bay,
watermen still speak Guinean, a dialect full of "ye" and other archaic words along with the lilt of dropped r's and d's of the greater
Tidewater area.

Gloucester County's population of 36,886 is spread over 217
square miles, with a median household income of $62,000. His-

---

* Correction: The original version of this article classified more of these disasters
as weather-related; as one reader pointed out, four of them were instead earth-
quake-related.

torically, most in the region relied on fishing and agriculture, but now many commute over the George P. Coleman Memorial Bridge to Williamsburg and Norfolk. A few McMansions have popped up, but they look out of place; I saw one fancy house sitting next to a trailer with a goat roped to the front porch.

Gloucester is one of the biggest losers in the geographic lottery when it comes to sea-level rise—low, surrounded by water, and flat as a billiard table. In 2003 Hurricane Isabel walloped the county, causing $1.9 billion in damage throughout the state of Virginia. And that was just the beginning.

On a warmer-than-it-should-be spring morning, I meet Chris West, a burly tugboat engineer whose gray rancher sits about 120 yards from the Severn River and about 3 feet above sea level. He's in his driveway, waiting for a team of contractors to jack his home 6 feet up in the air. West's parents built the house sixty years ago, and he's lived there his whole life.

West, forty-three, watches as the contractors clear brush and unhook his utilities; soon they will break the foundation, shove giant wooden beams under the house, and crank it up on hydraulic jacks. Then they will stack four wooden piers beneath the structure, like Lincoln Logs, to hold it up as they pour cement for a new foundation 6 feet higher. West and his dogs will stay at his girlfriend's house while the house is under construction.

"I got butterflies," he says.

Raising up the house would have cost West $85,000, but it's being subsidized by the Federal Emergency Management Agency's Hazard Mitigation Grant Program, so he only has to cover around $4,000. So far FEMA has given Gloucester County $11.8 million to raise sixty homes, and another fifty-nine homes and businesses are in line for a lift. The county also offers residents an outright buyout, but only seventeen families and one business have gone that route.

Neither FEMA nor county officials promote the grants as a climate change adaptation program—even if that's exactly what they are becoming. "Even the naysayers have noticed the increased number of storms," says Anne Ducey-Ortiz, Gloucester's director of planning. "Even when people don't want to deal with climate change, they are willing to talk about increased flooding."

For West, the reality hit home during Hurricane Isabel. The storm pushed water from the Chesapeake Bay into the Mobjack

Bay, then up into the Severn, through his backyard, over the deck, and into the house. He had 18 inches of seawater inside and nearly $30,000 in damage. He spent three and a half months living with his in-laws as he tore out drywall and carpets and replaced all his furniture.

Isabel was the worst storm in this region as far back as most people here can remember, and folks assumed the next one wouldn't come for a long while. "You think, 'Thirty more years I won't be living here anyhow, I'll be in a nursing home,'" West says. But then just three years later, a tropical storm named Ernesto blew through, bringing the water an inch and a half short of his deck. Then there was Nor'Ida in November 2009, and Hurricane Irene in August 2011, again bringing the water under the deck and inches below the threshold of his back door. Even the average tides behind the house, West reckons, seem almost a foot and a half higher than they used to be.

Tide measurements have found that the sea level along this part of Virginia's southern coast has risen 14.5 inches in the past eighty years, and scientists expect the rate of increase to double for the rest of this century, adding another 27.2 inches. Meanwhile, the land is getting lower: the earth here has been settling, due to the double whammy of a glacial retreat in the Pleistocene era 20,000 years ago and a giant meteor that hit some 35 million years before that. Removing massive amounts of groundwater for paper mills and other industrial uses has aggravated the sinking, much like the aquifer depletion that's been causing killer sink-holes in Florida. The upshot, says Pam Mason, a senior coastal management scientist at the Virginia Institute of Marine Science in Gloucester: "You take our little two-foot tide and you put one more foot on it, and it starts to make a difference. We've gotten complacent. We've gotten really close to the edge."

Ducey-Ortiz says that in a number of areas the county would prefer just to buy out homeowners because, even if you lift up the houses, flooding will submerge the roads, trapping residents and cutting off emergency services. Still, authorities won't force any-one to move. "You're allowing people to stay in a hazard area," she acknowledges. But "in Gloucester, that's our heritage, that whole Guinea area. To abandon those people, those families that live out there, people who just love living on the water—we want to help them."

West says he might have considered taking a buyout, but it would have had to be at least $200,000—what he owes on his mortgage right now. Like most in Gloucester, he elected to stay. "It's hard to take people out of their home, their true home," he says. Most of his neighbors, "the only time they've left has been in a pine box."

Perhaps the only topic touchier than whether people should abandon their homes is why the problem even exists. West has heard of global warming, but he's not entirely sure it's responsible for the rising water. "Nobody knows, I don't think. Everybody speculates," he says. Local authorities rarely, if ever, speak the words "climate change."

"I wouldn't want to turn some positive influences off by coming up with a political term," said Paul Koll, Gloucester County's building official. "I am really conscious of not labeling it anything so I don't shoot myself in the foot." Two years ago, when leaders in neighboring Mathews County broached the subject of sea-level rise, Tea Partiers packed meetings, warning of an environmentalist plot to "put nature above man." They linked a proposal to build dikes to a United Nations sustainability plan known as Agenda 21, which has inspired a number of conspiracy theories among far-right activists.

Never mind that the Middle Peninsula, made up of Gloucester and five other counties, expects up to $249 million in new climate-related costs by 2050, a figure that doesn't even include potential damage from storms like Isabel or Ida. The American Security Project, a Washington think tank, projects that climate impacts could cost the state a whopping $45.4 billion by 2050, with extreme storms alone putting $129.7 billion worth of property at risk.

Yet the Republican governor, Bob McDonnell, phased out a Governor's Commission on Climate Change after taking office in 2010. His attorney general, Ken Cuccinelli (who won the state's Republican gubernatorial nomination in May), has spent a good deal of his time in office seeking to prosecute the former University of Virginia climate scientist Michael Mann for his work on historic temperature records. And when state lawmakers requested a study on sea-level rise, the Republican state delegate Chris Stolle retorted that the term was "left-wing." (The legislature settled on "recurrent coastal flooding.") And Virginia is better off than neighboring North Carolina, where lawmakers last year explicitly refused to consider scientists' current projections in coastal building decisions.

Just as New York City planners had data showing that a superstorm could devastate the city, the federal government has plenty of data on the climate cliff—the looming budgetary catastrophe from emergency spending. In January the National Climate Assessment and Development Advisory Committee released a draft of its latest report, warning of the "high vulnerability of coasts to climate change." The report optimistically added that "proactively managing the risks will reduce costs over time." But with congressional Republicans actively derisive of climate science, the odds of that are not great.

The closest thing to a federal effort to mitigate climate risk may be the twenty-five-year-old FEMA grant program that pays for the house raisings in Virginia. But most of the money is designed to kick in *after* a disaster, to prevent recurrence—and it doesn't take into account whether houses in the floodplain are viable in the long term. Still, it's more proactive than the lion's share of FEMA's post-disaster spending, which allows people to rebuild in high-risk areas as if a storm or flood will never happen again.

I visited Representative Earl Blumenauer (D-Oregon) in his office the day after the $60 billion Sandy relief package passed the House, nearly three months after the storm. He wasn't happy with it. "What Sandy illustrated is both the increasing vulnerability—and the costs and consequences," he told me. "But we really didn't condition the recovery on making sure that we minimize people going back in harm's way."

A slight, bookish lawmaker whose lapel often sports a bright plastic bike pin, Blumenauer has been Congress's loudest critic of disaster policy. Even the Sandy package, he notes, had no incentives to consider climate change as part of rebuilding plans. If Blumenauer had his way, the federal government wouldn't rebuild any of its facilities in the floodplain—no post offices, no office buildings. Counties that get disaster relief would be required to enforce better building and zoning codes. And the feds wouldn't pay to keep rebuilding homes exactly as they were before a storm.

"In the aftermath of a disaster, the instinct is to reach out to people, to try to help them," he adds. "And so many people, their first instinct is to just go right back. It is devastating to look at all the things we do that keep people anchored in very dangerous places."

True, he says, it's hard to challenge people's yearning to rebuild. But at some point, lawmakers need to start thinking about the next cataclysm. "Before the next big wildfire, before part of the coast washes away, before the predictable unpredictable storm hits, what are the principles we're going to have?" Blumenauer asks. "What is going to be the condition of federal assistance?" With that in mind, he has been trying to expand mitigation programs for years, with limited success. Under the 1988 Stafford Act, 15 percent of the emergency relief funding Congress allocates to FEMA must be used for mitigation, but that money is allocated only after disaster strikes.

I asked Blumenauer if it's even practical, in the long run, to raise or relocate all of the homes that need it. "It's expensive," he says, "but it's a fraction of what we're routinely spending."

Why, then, do we keep spending more? One reason is that most disaster spending doesn't actually show up in the federal budget; it's treated as "emergency spending," which isn't included in regular appropriations. So while fiscal hawks in Congress leave disaster-preparedness programs chronically underfunded, disaster relief is treated as a budget freebie. The obsession with deficit reduction —codified in the 2011 Budget Control Act, which capped federal spending as part of the debt ceiling debate—has reinforced that trend. "You lowball it so you can spend the money elsewhere, but then you come in with the disaster supplemental, which is free money," Blumenauer says. Congress has had to pencil in extra funding for FEMA's Disaster Relief Fund in eight of the last ten years, after the appropriated amount fell short of the actual need. And it keeps happening; the Obama administration's proposed 2013 budget, for example, shaved 3 percent—$1 billion—off the Disaster Relief Fund.

Chronically underfunding disaster spending is shortsighted in the extreme, says Blumenauer. "We're just cannibalizing programs," he says. "We save arguably five dollars for each dollar we invest."

Just as disaster-relief funding ignores the fact that today's disasters are tomorrow's normal, the NFIP is virtually designed to ignore dramatic changes in weather and flood patterns. Created in 1968, it was made to help people in the most flood-prone areas acquire insurance—policies that private insurers were not willing to underwrite. In 1973 flood insurance was made mandatory for anyone who had a federally backed mortgage in an area considered at risk for a hundred-year flood; those already living in those areas were grandfathered in at heavily subsidized rates. Today the Property Casualty Insurers Association of America estimates that homeowners covered by federal flood insurance pay just half of the "true-risk cost" to insure their properties. In the highest-risk areas, they pay just a third.

No surprise, then, that the federal insurance program is now $25 billion in the hole. In the Sandy supplemental spending bill, Congress upped the program's borrowing authority by $9.7 billion, to $30.4 billion, to meet new claims—money that is unlikely ever to be paid back. And because the subsidy is so great, there's no incentive for private insurers to enter the market, says Frank Nutter, president of the trade group Reinsurance Association of America. "If you had a program that is fiscally sound, you'd probably find more insurance companies willing to write the risk," he says. "They wouldn't be competing with a government program that is underpricing the risk."

The other problem with subsidized insurance is that it encourages people to build—and stay—in high-risk areas, since they'll be bailed out even if they incur disaster after disaster. It's what economists call a moral hazard, a circumstance that encourages people to engage in risky behavior because the costs are borne by others.

"In many cases," says water consultant Conrad, "we've removed the most important element in our marketplace that forces the responsibility on the decision maker him- or herself." Conrad has been documenting the flood insurance program's problems since 1992. In 1998 he found that "repetitive loss" properties—those that had more than $1,000 worth of damages from a storm two or more times in a ten-year period—made up 2 percent of insured properties but were responsible for 40 percent of what the program was paying out. For one in ten of those properties, the program had paid claims that totaled more than the house's market value. (In response, in 1999 Blumenauer introduced the Two Floods and You're Out of the Taxpayers' Pocket Act, which never made it out of committee.)

The NFIP has long been a sore spot for both environmentalists and small-government conservatives. "It has been a very long-term subsidy for development in places where we simply shouldn't be developing at all," says Eli Lehrer, president of R Street, a libertarian think tank based in Washington. "There are areas that we've developed that probably shouldn't have been developed in the first place, and wouldn't have been if we had private insurance, or even actuarial rates in a public program." But reform has been tough—because every attempt at change meets firm opposition from politicians representing flood-prone districts and from local governments that rely on coastal properties for property taxes and economic development. "Every time they'd try to raise the rate, there would be a roar from up on Capitol Hill," says Conrad.

In 2004 Blumenauer did push through a major overhaul of the insurance program, including incentives to raise or buy out houses that had been damaged multiple times. But it took hurricanes Rita and Katrina, and a more deficit-minded Congress, to pass another flood insurance reform bill last year that finally limited subsidies for second homes and for properties that were damaged repeatedly.

Under that 2012 reform, such homes will see premiums rise dramatically over the next five years, eventually bringing 400,000 of the most heavily subsidized properties up to market rates. The new law also lets FEMA buy homes that are considered "severe repetitive losses" at their full pre-disaster price, rather than the 75 percent it offered before.

But perhaps the most significant change in the reform involved maps—specifically, FEMA's floodplain maps, which determine who must buy flood insurance. Those maps can now for the first time include "future changes in sea levels, precipitation, and intensity of hurricanes." But there's a catch: those changes don't affect the new flood maps FEMA is currently releasing, the first in thirty years. Floodplain maps issued for New York City and coastal New Jersey in late 2012 and early 2013, for example, don't account for sea-level rise. Maps for the rest of the country are rolling out slowly, and it's unclear when they will start including sea-level projections.

Back during the Bush administration, in 2007, FEMA began a major assessment of how climate change would affect the flood insurance program, with a projected completion date of 2010. When FEMA finally released the report in June 2013, it included a number of alarming findings. Rising seas and severe weather are expected to increase the area of the United States at risk of floods by up to 45 percent by 2100, doubling the number of people insured by an already insolvent program.

It took three and a half months to put Chris West's house up on a higher foundation. When Hurricane Sandy glanced off the Virginia coast just a few months after the contractors were done, everything at his end of the neck flooded, but the water flowed right underneath his house. "I don't have that worry anymore," West says, "of being displaced out of my home." A couple of other homeowners decided that they just wanted to leave; as of May, Gloucester County had acquired eighteen parcels of land, but since then eight more homeowners have signed up for buyouts.

As vulnerable as it is to rising seas, Gloucester is lucky. It has forward-thinking local officials who acknowledge the problem, even if they'd prefer not to talk about the root causes. Gloucester County has earned improving marks from FEMA for trying to minimize flood risks with steps like establishing tougher building codes and requiring all new development to be built at least 2 feet above flood height. Those initiatives lower the cost of insurance for homeowners, but they also save the federal government money —an estimated four dollars in future savings on property damage alone for every dollar spent on prevention.

Skip Stiles hopes an appeal to fiscal sanity is what will finally get decision makers to care about climate. Stiles, sixty-three, is the director of Wetlands Watch, a Norfolk-based advocacy group that formed back in 1999 to protect shoreline habitats. Not long after joining the group in 2005, Stiles realized that saving tiny parcels of marsh wasn't going to help much if the entire coast was wiped out by century's end. "We started realizing it's not just the wetlands —it's the whole freaking economy in this region that's at risk," he says.

That, and not that many people care about wetlands. "We said, 'What *do* people care about?' Their homes, their business, their way of life."

Stiles took me on a ride through Norfolk, highlighting spots that have seen major flooding in the past few years. He pointed to one house where a car floated into the front door during a storm, and another that the owner, tired of dealing with the water, has been trying to sell for months. We drove through the Old Dominion University campus, where a small permanent lake has formed in the back corner of a huge parking lot. "You can't pave under water," he noted dryly, "so this obviously wasn't under water when this parking lot was paved."

Stiles left Washington for coastal Virginia after twenty-two years as chief of staff to the late California Democratic representative George Brown, who in 1978 launched the first federal climate-change research program. But it was not until Stiles saw Norfolk's frequent flooding that he realized climate change, far from a distant threat, was a disaster well under way. "I thought, 'Oh, the feds are going to fix this,'" he says. "BS, ain't happening. It's local government—and man, the politics at the local level."

So Stiles started showing up at local planning commission meetings, begging officials to stop approving new shoreline developments in the face of inevitable sea-level rise. Back when he began, in 2006, "they looked at us like we were crazy"—coastal land is often the most valuable in a county, and it generates the highest property taxes. But then he stopped talking about climate change and started talking about budgets. "Suddenly it was like a key in the lock," he says. "What quickly happens is the money you put into those neighborhoods exceeds the property tax you get out. These neighborhoods turn into money pits. There just isn't enough money to raise all of the structures that need to be raised."

For Stiles, it doesn't matter whether local officials will actually utter the words "climate change," so long as they start dealing with the impacts. "Our perspective is just, 'Look, get on the bus, and we'll figure out together where the destination is,'" he says.

It's all about good, old-fashioned fiscal conservatism, says Conrad, the water consultant: "If this is all done by just pots of money being thrown around, most of the residents will be inclined to just take the money, do what's immediately convenient, and ignore the elephant in the room, which is that the Atlantic Ocean wants to move inland."

Matthias Ruth, an economist at Northeastern University who focuses on climate impacts, says the key is to provide a financial return for planning ahead. "We know that what once was the hundred-year floodplain now is the ten- or five-year floodplain. So what we need to do is give the incentives to either fortify buildings, elevate them, flood-proof them, or have a controlled retreat." Instead "we pretend it's not an issue, and we put ever more infrastructure into the coasts and ever more people."

Ruth ticks off the expected costs of climate change on the coasts—seawalls, flood insurance claims, disaster response, not to mention dislocation, stress on communities, and lost social connections. And what if, after a major storm like Sandy, there were an ice storm or maybe another hurricane the following year? "It's these one-two punches, the cumulative effect that they have on our infrastructure, our social systems," he says. "What we already see is worrisome, but this is going to be an order of magnitude more worrisome."

And with every year that goes by without shifting the incentives, both the costs and the future fiscal liabilities get larger. Many observers believe that after a major disaster, particularly one of Sandy's size and scope, there's a window—maybe six months, maybe a year—for a real shift in direction. Even with Congress frozen in denial, there's a lot the Obama administration could do: the Veterans Administration could stop underwriting mortgages on homes in flood-prone areas, and the Department of Commerce could deny economic development grants to projects on the coast. The Department of Housing and Urban Development could situate new low-income housing away from flood zones, and the Department of Transportation could build roads where they won't be under water in the near future.

We're already spending billions on responding to storms and disasters made worse by climate change, notes Ruth; Sandy gave us a chance to think differently. "Why don't we take [that money] and invest in infrastructure in ways to overcome the existing inefficiencies and improve quality of life?" he asks. "And then as we do this, reduce the vulnerability. Instead of having this downward spiral, have an upward one."

In the end, says Stiles, it might be a matter of how many disasters it takes to generate momentum. "I look at these little moments, this incremental progress, but I wonder, 'Is there enough time? Can we make it?'" he says. "Are there enough of these events coming up, and are we smart enough to catch up with the change that nature is going through?"

BILL SHERWONIT

Twelve Ways of Viewing Alaska's Wild, White Sheep

FROM *Anchorage Press*

1.

The image is a sentimental favorite, a portrait of two wild sheep that's suitable for framing or getting published in a book. A Dall sheep ewe and her lamb gaze directly at the viewer, only their white upper bodies and heads visible behind gray, lichen-splattered rocks. The faces of both appear calm. Inquisitive. Yet their large golden eyes, erect ears, and pursed lips also suggest caution. And maybe some uncertainty. There's a sense that the sheep will bound away if the human they're intently watching steps any closer or makes some sudden, awkward move.

The ewe-lamb pair were captured in this picture while perched on a steep hillside that looms above the Seward Highway south of Anchorage. They were among a half-dozen or so sheep visible from the highway that day, yet high enough on the cliff face and far enough from Windy Corner—a place known for its sheep-viewing opportunities—that most travelers missed their presence.

Unlike those who rushed past on their way to farther destinations, I'd come looking for these animals, which I knew to be highly tolerant of people. Armed with a high-quality camera, a newly purchased telephoto lens, and a burning desire to get some close-up wildlife shots, I painstakingly ascended the steep, wooded and rocky slope, hopeful that the sheep would allow me near them.

Though I'd come to Anchorage in the early 1980s to work as a sportswriter, I had other long-term (and at that point largely unspoken) ambitions: to someday be the newspaper's outdoors writer and to supplement that writing with my own photography.

Eventually I would delve into the nature, the life stories of Alaska's Dall sheep. But on this day I was less interested in observing their behaviors and habits than sharing their company and, especially, getting their pictures. In that regard, the day proved a heady success. A mix of mature ewes, adolescents, and lambs, the sheep allowed me to briefly join their company. At times we came within 15 or 20 feet of each other, their curiosity a match for mine, or so it seemed. Perhaps they wondered what foolish sort of human would risk his life to walk and stumble along such steep and crumbly slopes.

2.

The Brooks Range is where I saw my first wild sheep, while working as a geologist in the mid-1970s. During one daylong traverse I made in bright sunshine and energy-sapping heat, two short-spiked ewes appeared atop a ledge, like some far north mirage. Less then 50 feet above me, the sheep seemed more inquisitive than wary when I slowly passed beneath them. And why not? Deep in the Arctic wilderness, dozens of miles from the nearest road or village, I may have been the first human they'd ever met. The two continued to watch until I walked out of sight, my mood brightened and body enlivened by the surprise encounter with luminously white animals.

3.

It can be argued that Dall sheep, as much as 20,320-foot-high Denali (Mount McKinley), grizzly bears, wolves, or vast tundra expanses, are perfect symbols of what the famed biologist Adolph Murie called Denali National Park's "wilderness spirit." The snow-white mountain sheep are what drew naturalist-hunter-author Charles Sheldon in 1906. And their preservation, as much as anything, inspired him to seek park status for this wildlife-rich part of Alaska, a quest that led to the creation of Alaska's first national park—then named Mount McKinley—in 1917.

Later in the park's history, severe Dall sheep declines in the 1930s and '40s caused great alarm and forced park officials to confront wildlife-management policies that favored one species (sheep) over another (wolves). Thanks largely to Murie, the sheep crisis—and the species' eventual recovery—ultimately led to a strengthening of ecosystem rather than favored-game management in Denali. Here all native species would be protected.

Nowadays, an estimated 2,500 Dall sheep inhabit Denali National Park and Preserve's alpine heights; based on 2008–2009 surveys and those done in the mid-1990s, park biologists believe sheep numbers to be "fairly stable." The large majority live on the park's northern side, in both the Alaska Range and the Outer Range.

Because the Denali Park Road borders some of the sheep's prime habitat, visitors have an excellent chance of seeing these animals, though usually from a distance of several hundred yards to a half-mile or more. They are most often seen as tiny white dots on the upper flanks of tundra-covered hills, though occasionally they can be spotted near or even on the road. Visitors willing to climb hills are more likely to see the sheep up close, though park regulations prohibit people from approaching closer than 75 feet.

A small percentage of Denali's Dall sheep actually cross the park road on seasonal journeys between the park's two mountain ranges. These migratory sheep spend their winters in the Outer Range, where snowfall is light and high winds often keep exposed ridges free of snow. In May or June they form groups of up to seventy animals and cross wide lowlands to reach the Alaska Range's northern foothills for summer's green-up. There the sheep remain until autumn, when they retrace their steps. Biologists since Adolph Murie have known that a portion of Denali's sheep migrate, but it remains a mystery why some do and others don't.

4.

With a nod of thanks to all of those who've studied *Ovis dalli* and shared what they have learned, I'll now present a potpourri of natural-history facts about the white, wild sheep, which is named after the American naturalist William Healy Dall and inhabits mountain ranges throughout much of inland Alaska and neighboring western Canada.

Adult male and female members of the species live apart except during the early winter mating season in November and December. Just prior to the rut (and occasionally throughout the year), mature rams butt heads in fierce battles that scientists say determines their place in the band's social order and, consequently, its breeding order. Facing each other, two rams rear up on their hind legs, then charge and "clash horns" with a loud bang that's been compared to that of a baseball bat slammed into a barn door. Adult females too will sometimes knock heads, apparently to determine social rankings.

Ewes produce a single lamb in late May or early June. As the birth approaches, a pregnant ewe will go off by itself and head for steep, rugged terrain where predators are less likely to be. Lambs usually do fine their first summer, when food is abundant, but half or more may die their first winter. Sheep that survive their first couple of winters may live into their teens, though biologists consider twelve to be "very old" for a wild sheep. Mature rams in their prime may weigh 200 pounds or more, ewes 110 to 130 pounds on average.

Both sexes of adult sheep have horns, though only males grow the large, sweeping, and outward-curling horns so often seen in photos. As rams mature, their horns gradually form a circle when viewed from the side and reach a full circle or "curl" in seven to eight years. The amber-colored horns are male status symbols; large mature rams can sometimes be seen displaying their horns to other sheep as a sign of their dominance. Those of females are shorter, slender spikes that resemble the horns of mountain goats, which sometimes causes people to confuse the two species. But goat horns are shiny black and sharper than sheep horns, and goats also have more massive chests. Besides that, their ranges rarely overlap, goats being most common in coastal mountains and sheep inland.

Unlike the antlers of moose and caribou, horns are never shed. Their growth occurs only from spring through fall; winters are marked by a narrow ridge or ring. So, much like a tree's age, the age of a sheep can be determined by counting its "annual rings," also called annuli.

Dall sheep are grazing animals that feed on a variety of plants, including grasses, sedges, willows, and herbaceous plants; in winter they survive on lichens, moss, and dried or frozen grass. They prefer to stay up high, in places that combine open alpine ridges and meadows with steep slopes, because their hill-climbing skills make it easier to escape predators in such sheer, rugged, mountainous terrain. Wolves are the most efficient predators of sheep, but grizzlies, coyotes, lynx, and wolverines sometimes successfully hunt the species, and golden eagles prey on young lambs.

5.

In the early 1900s, when prospectors and pioneering mountaineers were lured into the Denali region by gold and summit fever, respectively, an easterner, Charles Sheldon, came north to the Alaska Range on a different sort of quest: a big-game hunt. What Sheldon found in 1906 proved far more valuable than any trophy animal. Denali's wilderness and wildlife sparked the idea for a park-refuge unlike any other in the nation.

By all accounts a skilled hunter, passionate naturalist, devout conservationist, successful businessman, gifted writer, and astute political lobbyist, the Vermont-born and Yale-educated Sheldon was passionate about all species of mountain sheep, which he believed were the noblest of wild animals. He studied and hunted them throughout their North American range, finally pursuing his passion to the most remote part of the continent, in search of Alaska's fabled white sheep. For a guide he chose Harry Karstens, a transplanted Midwesterner known to be a first-rate explorer (Karstens would later become the first superintendent of Mount McKinley National Park).

Accompanied by a horse packer, Sheldon and Karstens lived off the land while they explored the Alaska Range's northern slopes. By mid-August they had seen hundreds of sheep, but all were ewes, lambs, adolescents, or young adults. Finally, while hiking Cathedral Mountain alone on August 17, Sheldon discovered a band of mature rams, including nine with "strikingly big horns." In great and dramatic detail, he described his solo stalk of those "big rams!" in *The Wilderness of Denali*. Here I'll share a short excerpt that recounts the climax of his hunt:

Finding a slight depression at the edge [of a canyon that separated him from the sheep] I crept into it and lay on my back. Then slowly revolving to a position with my feet forward, I waited a few moments to steady my nerves. My two-hundred-yard sight had been pushed up, and watching my opportunity, I slowly rose to a sitting position, elbows on knees. Not a ram had seen or suspected me. I carefully aimed at a ram standing broadside near the edge of the canyon, realizing that the success of my long arduous trip would be determined the next moment. I pulled the trigger and as the shot echoed from the rocky walls, the ram fell and tried to rise, but could not. His back was broken. The others sprang into alert attitudes and looked in all directions. I fired at another standing on the brink, apparently looking directly at me. At the shot he fell and rolled into the canyon. Then a ram with big massive wrinkled horns dashed out from the band and, heading in my direction, ran down into the canyon. The others immediately followed, but one paused at the brink and, as I fired, dropped and rolled below.

A few hundred more words follow, describing Sheldon's continued killing spree. Then this:

Seven fine rams had been killed by eight shots — and by one who is an indifferent marksman! My trip had quickly turned from disappointment to success.

The U.S. Biological Survey had entrusted me with the mission of securing the skulls of at least four adult rams, with some of their skins, for the study collection in the National Museum, and I desired four reasonably good trophies (the legal limit) for myself. Most of these were now before me.

The rain had stopped. I sat there smoking my pipe, enjoying the exhilaration following the stalk, while the beauty of the landscape about me was intensified by my wrought-up senses.

6.

Given what I knew of Charles Sheldon — legendary champion of Alaska's wild sheep and Denali's wilderness — his account of the killing spree stunned me when I first read it.

I understand the unfairness of using contemporary standards to judge people who lived in other eras, under different value systems and moral codes. In his time, Sheldon was considered a consummate conservationist. And by most accounts, he worked harder than anyone to get the homeland of these sheep protected. He is celebrated for his wilderness advocacy, especially his role in getting the federal government to establish Mount McKinley National Park, later to become Denali National Park and Preserve. Still, both the actions and the attitude he exhibited that long-ago day disturb me.

Except for catching and killing fish (and I don't do much of even that these days), I am what's called a nonhunter. I don't directly kill other animals for food, or clothing, or other reasons. But contrary to what some of my Alaskan critics say, I am not an "anti." I do not oppose the respectful and humane hunting and killing of animals for food or other subsistence purposes. It seems more honest, in a way, and arguably more ethical than buying meat at the store, especially given what we know about the awful lives of most farm animals that become our food. But as I've noted elsewhere, over the years I've become intolerant of trophy hunting. To me such blood sport is unacceptable, a selfish and harmful act fed by pride and ego.

Can I accept Sheldon's behavior? Given the context, I suppose. Yet I'm bothered by his excitement and self-congratulatory tone (I read false humility in the comments about his marksmanship) and especially his cavalier attitude toward the sheep he killed. Sheldon suggests no sense of regret or sadness that he took their lives. Yes, he obeyed the hunting laws. But to kill seven rams for science and personal satisfaction seems the epitome of overkill, no matter how healthy the local sheep population may have been.

End of commentary, back to Sheldon's story.

7.

To his credit, Sheldon painstakingly recovered the animals he'd killed. One by one he butchered the sheep and hauled their meat, skins, and skulls down the mountain, then treated them for preservation, took measurements, and studied the stomach contents. Once reunited with his companions, he packed his specimens out and returned east. His trip had been a resounding success.

Sheldon's main work was now complete. Several of the sheep he'd killed and collected would be studied by scientists and displayed in the American Museum of Natural History. But he recognized the need for a longer stay, to better understand the sheep's life history. So before leaving, he built a cabin along the Toklat River.

The following August, Sheldon returned for a ten-month stay. Besides studying sheep, he gathered facts on other species, large and small. He paid close attention to the landscape, the wildlife habitat, and the changing weather and seasons, and he made friends with many year-round residents. He was astounded by the abundant wildlife, but feared that the uncontrolled hunting done by "market hunters" who supplied wild meat to the region's towns and mining camps would someday threaten Denali's wildlife riches.

Along the way, Sheldon fell in love with the Denali region and began to envision a plan for its preservation as a park and game preserve. In a journal entry dated January 12, 1908, he even named this park-refuge: Denali National Park. Though he never again returned to Denali country, it remained an inspiration for this dream.

Back in New York, Sheldon shared his Denali vision with fellow members of the Boone and Crockett Club, an influential group of big-game hunters and conservationists. His colleagues responded enthusiastically, but momentum for the new park-refuge lagged until 1914, when Congress mandated construction of a railroad from Seward to Fairbanks. The preferred route would pass through the heart of the Denali region.

Sheldon pursued his dream with new urgency. Joined by equally enthusiastic allies—the newly appointed Park Service director Stephen Mather and the prominent artist-explorer-hunter Belmore Browne among them—Sheldon helped draft legislation establishing Mount McKinley National Park. He also won the backing of Alaska's lone delegate to Congress, James Wickersham, who realized that such a park could be a major tourist attraction, boosting the territory's economy.

With new momentum and little opposition, Congress passed legislation creating the park in February 1917. Later that month, President Woodrow Wilson signed it into law. The original Mount McKinley Park protected 2,200 square miles of prime wildlife habitat, primarily north of the Alaska Range, where Sheldon had met the Dall sheep that would help to define his life and conservation legacy.

8.

Traffic slows to a standstill. Cars, pickups, and RVs begin pulling over to the side of the highway. Binoculars and cameras (and, by the second decade of the twenty-first century, smartphones) are grabbed. And a crowd begins to gather, as both tourists and Alaskans maneuver for a better look at the Dall sheep that feed less than 100 feet away.

The sheep pay little attention to the human spectators. Continuing to feed on grass and willows, they sometimes wander close to the road and show no outward signs of fear, even when people approach to within 30 feet or less. It's a scene that's repeated dozens of times each summer, along one of Alaska's busiest stretches of highway.

Tens of thousands of Dall sheep inhabit the state's mountain ranges, from Southcentral Alaska to the Arctic. They're prized wildlife symbols of three national parks: Denali, Wrangell–St. Elias, and Gates of the Arctic. But nowhere are they so accessible to the public as the Windy Corner area of Chugach State Park, a half-hour's drive from downtown Anchorage and the only place in the world that people can watch Dall sheep while both are standing near sea level.

Ewes, lambs, adolescents, and young adult sheep inhabit steep cliffs and grassy meadows above the Seward Highway for much of the year, coming closest to the road between mileposts 106 and 107. Peak viewing occurs in summer, after the ewes have given birth. The best time to see the sheep is usually early morning, though they're sometimes visible throughout the day. As many as fifty have been spotted from the highway, but twenty or fewer is more the norm. Only rarely are the older, big-curl rams present; they seem to prefer backcountry solitude to busy highway corridors.

While the sheep's high visibility is a guaranteed treat for wildlife lovers, it has proved a management headache for Chugach State Park personnel and state troopers. Drivers who slow down or stop

to watch and photograph the sheep may ignore designated turn-
outs and park instead along narrow highway shoulders, despite NO
PARKING signs. Or, even worse, they'll slow almost to a stop on the
highway itself. And as crowds gather, people pay less attention to
traffic patterns.

As a former Chugach superintendent, Al Meiners, once described it to me, "When people see wild sheep three feet from
the road, they just go nuts. Other senses tend to shut down and
you get people doing foolish things, like slamming on their brakes
right on the highway. It's real dangerous, because you have other
drivers coming screaming around that curve at fifty, sixty miles an
hour, and here's a traffic jam. I went down there once to study the
problem and ended up directing traffic."

State officials have talked for years about ways to better address the Seward Highway's "sheep jams." Some improvements
have been made, but the problems persist. Now I've learned that
a major highway redesign is in the works. Plans for Windy Corner
include a widened road with passing lanes, parking lots on both
sides of the highway, informational signs that present responsible
wildlife viewing behavior, a pedestrian tunnel, and perhaps some
sort of barrier to better separate people and sheep.

9.

A few years after becoming the *Anchorage Times*'s outdoors writer
in 1984, I met with Dave Harkness, the state biologist responsible
for managing the Anchorage area's wildlife, to learn more about
Windy Corner's sheep. Dave told me that biologists weren't absolutely certain why sheep would congregate in such large numbers
along a busy highway, but suspected a mineral lick, where they
were likely getting salt and other essential minerals from the soil
(those suspicions have since been confirmed). He also believed
that the cliffs contribute to their tolerance of human traffic: "The
sheep know they have an easy escape route if they need it. In a few
minutes, or even seconds, they can be out of view."

Harkness further noted that the Windy Corner sheep frequent
an area that is off-limits to hunting, which would help to account
for their "tame" behavior: lambs learn early that people don't pose
a threat. Yet, he added, "Come August and September, the sheep

are vastly different creatures; they're not as accessible or visible. It's hard to say whether they equate danger with different times of the year."

Neither Harkness's successor, Rick Sinnott, nor the Anchorage area's current wildlife manager, Jessy Coltrane, have noticed such a seasonal shift in the sheep's behavior. And in a conversation with me, Coltrane questioned why the animals would suddenly become more wary. I can offer one possibility: sport hunters.

The one obvious danger that Alaska's Dall sheep face in late summer and fall is human hunting. Of course it's difficult, if not impossible, to know whether sheep make that seasonal connection. But if Harkness was right, it's telling that they become "different creatures" when the killing season begins, or at least did so during his watch. Yes, sheep are protected at Windy Corner and on neighboring terrain. But parts of Chugach State Park's backcountry are open to sheep hunting from early August through early October, including one area less than 5 miles away, as a sheep rambles. Could the skittish behavior that Harkness observed be mere coincidence? I, for one, can imagine that sheep might somehow "learn" to associate that time of year with a need for greater caution.

On the other hand, it's curious that neither Sinnott nor Coltrane has witnessed such behavioral changes in the two decades or so since Harkness retired. And there's this to consider: as Coltrane points out, the great majority of Chugach hunters have always targeted full-curl rams. Why would the killing of rams make ewes and young sheep more cautious, especially since mature males generally keep to themselves?

Any permitted hunter could kill ewes in Chugach State Park from the mid-1990s into the 2000s, though since 2009 only archers have been allowed to take them. (Coltrane says that the hunting season begins after lambs have been weaned.) Between 2010 and 2012, hunters killed thirty-nine full-curl rams and only three ewes in the park and some adjacent lands. The paradox is that no ewes could be hunted when Harkness was manager, yet that's when Windy Corner's sheep apparently were most guarded during the hunting season. Could he have misread their actions? Could Sinnott and Coltrane simply have missed the seasonal shift? Or perhaps some other circumstance has changed. Such enigmas reflect how little we actually know about Dall sheep (and other animals), including—and especially—their inner lives and communication with each other.

A few additional observations about Chugach State Park's sheep and the hunting of them seem appropriate. More than two thousand sheep were annually counted from the late 1980s through the 1990s, about double their number in the early 1980s. Wildlife managers correctly suspected that was more than the sheep's habitat could indefinitely sustain, and Sinnott added new hunts in the mid-1990s to trim their numbers. The park's sheep population plummeted in the early 2000s, perhaps because of overgrazing and severe winters, though no one knows for certain. By 2007 about nine hundred sheep remained. Since then the population has rebounded slightly, to about a thousand animals.

Because, in Coltrane's words, "We micromanage the hell out of 14C [the unit that includes Chugach State Park]," managers are confident that the human kill didn't contribute in any substantial way to the sheep decline. And the allowable harvest has been lowered as sheep numbers dropped. For now the park's population seems stable.

10.

When I lived on Anchorage's Hillside, I could sometimes watch Dall sheep from my front yard. Even through binoculars, they were small white dots on the green or brown slopes below a Chugach landmark called Rusty Point. I always considered it a marvelous thing, to watch wild sheep move about their alpine homelands while I stood in my own suburban neighborhood, with its houses, roads, gardens, garbage pickup, and lawn mowers.

11.

Though my geology career ended in the 1970s, I've periodically returned to the Central Brooks Range in the years since then, as writer, adventurer, and wildlands advocate. Now largely protected by the 8.2-million-acre Gates of the Arctic National Park and Preserve, it has remained my favorite wilderness, a largely unpeopled landscape where one can spend days, even weeks, without seeing obvious signs of humans; a place where knife-edged ridges stretch in waves to the horizon and beyond, and glacially carved basins grow lush with tundra plants in summer; a place enriched by free-roaming grizzly bears, wolves, wolverines, sheep, and other northern animals.

In August 2008, I traveled to Gates on a ten-day solo trip. My visit got off to a rough start, with a broken tent pole and wintry weather, raising worries and dampening my spirits. My second day in the range was mostly rainy and gusty, so I stayed inside the tent except to cook and stretch muscles on a short hill climb. That hike proved to be exactly the tonic I needed. Step by step my worries washed away while I gently slipped into the wonder of those wild and ancient mountains. It was as if the harsh, rugged surroundings somehow eased me into a more tranquil state of being. The shift was subtle, until finally a thought hit me brightly: Wow, it's great to be back.

Near the end of my ascent, I happened to spot four sheep with short, spiked horns in a creek bottom below me. One looked small enough to be a lamb, its horns barely nubs. Unaware of me (as far as I could tell), they moved steadily yet leisurely up the valley, the smallest one prancing playfully at times among the boulders and meadows. Looking through binoculars, I then spotted three more sheep on a neighboring ridge.

Once atop the hill I hunkered down out of the wind and pointed my binoculars east, across a broad river valley. Scanning the distant hills, I first found a group of five sheep, resting on a dark slope. Then two more. And on another mountainside, nine sheep were scattered across the tundra. I was too far away to tell if any were mature males; I wondered about this, because all of the sheep across the valley were on national preserve lands, where trophy hunting of big-curled rams is allowed. By the end of my walk, I'd counted thirty-one sheep. As I noted in my journal, "Not a bad sheep-hunting day," at least for a guy armed only with binoculars.

A few days later, while hiking through a neighboring valley in designated wilderness, I made a curious discovery: a bleached ram's skull, its massive horns still attached. The horns nearly made a full curl, and both they and the skull were remarkably well preserved. What struck me as strange was the skull's position. It sat upright on a tundra bench, facing north, with no other skeletal remains, body parts, or hair nearby, as if someone—or something?

—had carried the skull to this spot and placed it carefully on the ground.

The horns' annular rings showed that the ram lived to be six years old, and I tried to imagine how one in its middle-aged prime might have died. Deep inside the park's wilderness boundary, he couldn't have been killed by a sport hunter, at least not legally. Park rules do allow residents of the area to lawfully hunt animals for subsistence purposes throughout Gates of the Arctic. But would such a hunter have left this beautiful head? And this locale seemed remote, even for someone aided by plane or snowmobile; there are many easier places to hunt.

Neither a grizzly nor wolves were likely to kill a ram during the peak of its life, unless the sheep was injured or otherwise infirm. Could a ram be so severely wounded during the rut's head-butting battles that it would become easy prey for a bear or wolf? Or starve? Maybe the animal died in an accident of some sort, an avalanche or a fall. Were there clues that I was overlooking? I reluctantly left the spot, filled with questions about the ram's life, its last days and hours.

12.

Nearly three decades have passed since I stalked the Falls Creek sheep on that steep hillside above the Seward Highway. Nowadays when I venture into the Chugach Mountains, I carry only a small point-and-shoot digital camera, or none at all. And when I watch Dall sheep it's usually from afar, through binoculars. Because of my own habits and favorite haunts, the place where I most often see them is along the rock-and-tundra-quilted flanks of Wolverine Peak, on a south-facing slope below the ridge that leads to Rusty Point.

Besides its southern exposure and abundance of rocky knobs and cliffs, that hillside must produce plenty of sheep food, because they can be found there throughout much of the year. One time I counted seventy-two animals scattered across the slope, a mix of ewes, lambs, adolescents, and young adults. More commonly I'll spot a dozen or two sheep. Now and then some large, full-curl rams (or nearly so) also congregate here, keeping to themselves as older males do.

Over the years, Wolverine Peak and Rusty Point have become two of my favorite hiking destinations, for any number of reasons: the terrain, the views, the easy access to high country, the exercise, the wildlife. For Rusty Point especially, nostalgia and solitude also draw me back. I still recall looking up there from my Hillside yard, watching those sheep. Over time, that distant rocky hilltop began to feel like an extension of my own neighborhood.

Though I've moved to another part of town, Rusty Point remains a special place to me. After reaching the point, I find a comfortable spot among the rocks and grasses and wildflowers, remove my pack and sweaty shirt, add layers of dry clothing, eat snacks and drink water, write in my journal, and take in the world immediately below me: mountains and city side by side, and then the landscape that stretches far beyond Anchorage.

Eventually I gather my belongings and follow the ridgeline, looking for signs of sheep. Almost always I'll find their scat—piles of small brown pellets—and clumps of coarse white fur. Occasionally I see the sheep themselves, grazing or resting on the slopes below. There is something wondrous and reassuring about this, to be up high in their rugged domain, to be walking where sheep walk and otherwise lead their wild and largely mysterious lives, which sometimes overlap with mine, however briefly.

The Separating Sickness

FROM *Harper's Magazine*

EDDIE BACON WAS A FORKLIFT OPERATOR at Trident Seafoods in Akutan, Alaska. In the summer of 1999, he developed mysterious rashes on his hands, arms, and legs. He visited a doctor, who gave him a variety of ointments, but they did nothing. He grew weak, lost weight. He had trouble seeing. No longer able to earn a living, he moved into his parents' house in central California. There, at a New Year's Eve party in 2000, he passed out, and his parents took him to the emergency room. He had green blisters on his hands, his weight had dropped to 90 pounds, and he couldn't stand up by himself. The medical staff at the hospital regarded Eddie with puzzlement and dread, asking his parents to put on gloves, masks, and gowns when visiting him. Finally, after four weeks, an infectious-disease specialist solved the mystery. Eddie had leprosy. Although a quarter million new cases of leprosy were diagnosed worldwide in 2011, only about 173 of those were in the United States; it's no surprise Eddie had a hard time getting a diagnosis. Once his symptoms were explained, though, his doctors could prescribe a venerable course of treatment: in June 2001, they sent him to the nation's largest leprosy clinic, which had recently relocated to Baton Rouge from its historic home in Carville, Louisiana.

The Louisiana Leper Home was founded in 1894. Infected residents were provided free room and board and medical care, but for the first fifty years of the home's existence, until the rules progressively relaxed in the late 1940s and early 1950s, they were denied the right to vote, to marry, to live with an uninfected spouse, or to leave the grounds. Children born to residents were taken away shortly after birth and put up for adoption. The Carville facility has been federally run since 1921, and the government allowed for the construction of a golf course and a lake, but there was also a cyclone fence topped with three rows of barbed wire and a jail to punish those who left the grounds without permission. In the 1950s, even after a cure for the disease had been found, escapees were still brought back in iron shackles.

Leprosy is really two diseases: the physical effects and the social response to them. In Hawaii, where leprosy was widespread in the nineteenth and twentieth centuries, it was called the "separating sickness." Once diagnosed, Hawaiian sufferers were hunted down like outlaws and offered a choice of exile or death. Those who chose exile were sent to a bleak camp built below the great cliff at Molokai (the leper colony there didn't close until 1969). And Hawaii wasn't alone. For centuries, from India to Iceland, people with leprosy—most don't like being called lepers—were ostracized. Only in the past sixty years have even a minority of leprosy patients received truly humane care.

The change started when doctors realized that leprosy, contrary to long-standing belief, is very nearly the least contagious contagious disease on earth: in more than a hundred years of Carville's operation, no employee ever caught leprosy from a patient. Ninety-five percent of us are naturally immune to the disease, and the rest have a hard time catching it. If you contracted leprosy at any point before 1941, your illness would not have been treatable, but your prognosis would not necessarily have been dire. If you were very lucky, your immune system might eradicate the bacterium on its own, or you might manifest only skin problems—lesions, eruptions, thickenings, numb spots—and little more. (Then as now, many with untreated leprosy live decades beyond diagnosis, into old age.) If you were less lucky, you might endure the collapse of the cartilage in your nose, an infection of the throat that could require a tracheotomy, or a permanent swelling of the face called leonine facies. Worst of all, you might suffer peripheral neuropathy—an absence of sensation that can cause patients to lose fingers, toes, and feet; blindness; and a host of other problems.

In 1941 a highly toxic tuberculosis drug, Promin, was shown to eradicate the bacterium that causes leprosy. Promin was tested on patients at the Leper Home, who improved dramatically; it was called the miracle at Carville. Painful Promin injections were fol-

lowed by dapsone pills, and by 1981 an even more effective mul-
tiple-drug solution was found. The disease is universally curable
now with a year or two of treatment. If you are diagnosed early in
countries such as the United States, you're likely to be disease free
and left with no evident disfigurement or damage. The problems
of leprosy now lie elsewhere—in the lasting stigma against sufferers, in the lack of resources for treatment in the developing world,
and in the very rarity of the illness in the developed world, where
doctors may not diagnose it in time to prevent permanent harm.

Eddie Bacon was unlucky. He had no natural immunity. But he
was also lucky: he was sent to Louisiana at a time when he could
be treated for both the bacteriological and the social effects of his
disease. After his initial treatment, Eddie returned to Baton Rouge
regularly for therapy to teach him how to take care of his body,
which has suffered some irreversible damage.

"It was a nice place," he recalled when I spoke with him in
2010. "I was scared they were going to treat me bad, but when I
went there, those people were touching me, they were hugging
me, those doctors and nurses." That physicality made him feel less
like an outcast. At the clinic in Baton Rouge, he said, "it's like I
have a second family." Eddie told me he's known as a "miracle
guy" because he survived one of the worst modern cases of leprosy
recorded in the United States, one that struck harder and moved
faster than the disease usually does. He's had to have both feet amputated, has scarring on his hands and arms, lost an eye (it went
blind and scared his nieces, so he had it replaced with a prosthe-
sis), and, like many others with the disease, lost his eyebrows. But
he is alive and he is free—and in December 2010 he got married.

To get to Carville, you drive northwest from New Orleans, leaving the highway after an hour for back roads through swampland,
where people live in trailers and old wooden shacks. Chemical-
manufacturing plants line the road, fenced-off palaces several stories high, wreathed in cold lights, giving off emissions that sharpen
the smell of the air. Past a few of these plants, the road dead-ends
at a T junction near a gas station, a Baptist church, and a cemetery
—all there is of the townless town of Carville. If you turn right and
drive along the high levee that hides the Mississippi, you arrive
after a couple of miles at what is now called the Gillis W. Long
National Guard facility, set on 350 acres of flat green land dotted

with handsome old oak trees. Near the entrance is a cluster of white buildings—a colonnaded plantation manor surrounded by houses, chapels, and industrial structures. A few patients still live at the center, and there is a museum of memorabilia from the days when hundreds of people lived and died here.

Many relics of the Leper Home remain: the covered walkways from building to building on which patients traveled on foot, on bike, and by wheelchair; the 1930s-era hospital that became a hotel (in which I was lodged); the cemetery. The museum displays ledgers and registry books; long, hand-knit bandages from the Ladies Auxiliary; antiquated medical equipment; a wheelchair made up as a Mardi Gras float; a pair of iron shackles; and numerous photographs of the residents at work and play.

On my first visit to the museum, in late 2009, I met a small, vibrant octogenarian with twisted, shortened fingers. Mr. Pete, as Simeon Peterson likes to be called, came to Carville as a patient in 1951, when he was twenty-three, and has lived here ever since. After years as a caretaker for those sicker than he, Peterson now greets visitors to the facility. He couldn't imagine leaving Carville. Where would he go? Most of his immediate family had died, and he was hardly ready to venture out into the world on his own.

It was lonely now, he told me. He thought the best time was his first decade as a patient. Back then, Carville had its own movie screenings, a newsletter, a baseball team, and enough patients to support a number of other clubs and activities. After the disease was rendered a fully treatable, outpatient disorder, the community dwindled. Most of those who remained were older patients damaged by the disease earlier in life and then by years of isolation.

A few decades ago, there were hundreds of patients at Carville, along with the nuns, medical staff, cooks, janitors, and others. When I first visited, sixteen residents remained; by the time I returned in early 2010 it was down to ten, and today there are six. The silence at Carville is profound. Only the occasional squadron of National Guard recruits marching and roaring in unison, the bugles for reveille, and the songbirds in the trees made any noise. As I wandered the grounds, I saw the name STANLEY STEIN on the sign for a U-shaped entry drive. He has his own exhibit in the museum, one that makes clear both the power of the community and the depth of isolation at Carville.

Stein worked as a pharmacist in Texas until, in 1929, he saw a doctor about some mysterious lesions he'd developed. The doctor reported him to the state board of public health. Although his symptoms were mild, he was locked up in a hospital isolation ward for several weeks, then packed, under escort, onto a train for New Orleans. An automobile took them the rest of the way, along the rutted road to the former plantation tucked into one of the Lower Mississippi's bends.*

Stanley Stein was the pseudonym that Sidney Levyson took upon entering Carville. "Have you decided on your new name, young man?" a nun named Sister Laura asked him when he arrived, in early 1930. "I was not just a sick man entering a hospital," Levyson recalled in his memoir, *Alone No Longer.* "I was a lost soul consigned to limbo, an outcast, and I must spare my family from any share in my disgrace." He stayed at Carville for thirty-seven years, the remainder of his life, starting a newsletter, the *66 Star,* that advocated for the rights of those with leprosy in the town and around the world, and he persuaded the government of Louisiana to repeal its forcible-quarantine law. Levyson also campaigned to have the name "leprosy" replaced with the less fraught "Hansen's disease," after the scientist who first identified *Mycobacterium leprae* —a change that has since caught on among scientists and doctors, if no one else. When he died, in 1967, he was buried not in Carville but in San Antonio, in his family plot.

Thomas Gillis, a researcher at the National Hansen's Disease Programs (NHDP) in Baton Rouge, told me he thinks of *M. leprae* as something of an invalid itself. "Most geneticists feel it's going through a degenerative process, losing parts it doesn't need," he said. The bacterium is delicate and slow. Because it doesn't cause symptoms until years or decades after it's contracted, patterns of transmission are still obscure. Unlike almost all other bacteria, *M. leprae* cannot be grown in the laboratory, putting ordinary research methods out of reach. Whereas some bacteria reproduce every twenty minutes, this one reproduces once every two weeks or so,

---

* Stein fared better than Mock Sen, a Chinese man who a few decades earlier had been imprisoned in a boxcar and shuttled between Baltimore and Philadelphia. Neither city would accept him, and so for thirteen days he was sent back and forth, until the boxcar was opened and he was found to have solved the problem by dying from exposure.

and it dies shortly after being taken from living tissue. You can get it to infect the hind feet of laboratory mice, but the only naturally susceptible animal is the nine-banded armadillo, whose low body temperature is an encouraging home for the fragile bacterium; as many as 10 percent of all the nine-banded armadillos in southern Louisiana and Gulf Coast Texas may be infected. Leprosy research in the United States involves paying trappers or sending out lab personnel to drive along the levees at night to scoop up armadillos in big nets. Dozens of the creatures reside in the armadillo quarters at the NHDP's research center, waddling along as naively oblivious as any vector could be.

Today, most leprosy research focuses not on the bacterium but on patient care. The disease can ravage the skin, but more devastating are its effects on the nerves, particularly those in the hands and feet, which it can cause to swell in their sheaths and strangle from lack of blood, leading to numbness. The eyes may be damaged if the nerves that cause blinking or those that relay pain and irritation fail. Without blinking, the eyes are not kept lubricated and clean; without pain, irritants are not felt, or a hand rubbing at the eye can scratch the cornea.

It used to be thought that tissue damage was a mysterious aspect of the disease itself, "bad flesh" that was somehow self-destructive, a belief so entrenched that the Wikipedia entry for leprosy featured the bizarre and unscientific phrase "auto-amputation," as though digits fell off like antlers or autumn leaves, until I changed it—twice. But in the 1950s, not long after Promin and then dapsone began changing everything at Carville, a doctor named Paul Brand noticed something unusual about the boys with leprosy he had treated in Vellore, India, while working as a missionary: patients were showing up with further injuries even after they had been cured.

Brand soon discovered that they were injuring their hands and feet in the course of everyday activities that the rest of us survive unscathed. A young carpenter he'd operated on, happy to be able to hold a hammer again, was oblivious to a splinter on the handle that ate into his palm all day. A farmer was hurt by his hoe, whose handle had a jutting nail. (At Carville, Stein once badly burned his hands in scalding water from the tap.) Feet, Brand found, were even more prone to injury: without the little adjustments we make in gait and shoe fit, sores arise and then become infected. And in a truly awful discovery, Brand and his Indian patients found that fingers and toes that inexplicably vanished overnight had been gnawed off by rats. From then on, the doctor gave each of his departing patients a cat.

For Brand, all this was revelatory: although his findings were grisly, they suggested that leprosy was far more treatable than previously thought, not only by bactericides but by training and educating patients on how to cope with the long-term effects of the disease. In his book *The Gift of Pain*, Brand writes about arriving at

> the theory that painlessness was the only real enemy. Leprosy merely silenced pain, and further damage came about as a side effect of painlessness . . . If we were right, the standard approach to leprosy treatment addressed only half the problem. Arresting the disease through sulfone drug treatments was not nearly enough; health workers also needed to alert leprosy patients to the hazards of a life without pain. We now understood why even a "burnt-out case" with no active bacilli continued to suffer disfigurement. Even after leprosy had been "cured," without proper training patients would continue to lose fingers and toes and other tissue, because that loss resulted from painlessness.

Brand went on to become the director of rehabilitation at Carville, where he worked from 1966 until his retirement in 1986. He continued his work as a surgeon there, pioneering tendon transfers that would restore function to "claw hands," a common effect of leprosy, but he also taught patients about the dangers of insensibility and the ways to compensate for it. He wrote that he felt like he was "introducing the boys to their own limbs, begging their minds to welcome the insensitive parts of their bodies . . . They lacked the basic instinct of self-protection that pain normally provides." Pain is what we feel when something has gone wrong, when our body has been neglected or mistreated. The people at NHDP today talk about "protective sensation," the feeling that allows healthy people to take care of themselves and warns them before sensation slides over into pain. Just as we relace our shoes before we get a blister, let alone an open wound, so we feel before we feel pain. The NHDP doctors teach their patients to take care of themselves with an empathy born of imagination and intention: to love their own now-alienated bodies as they might the body of someone else entrusted to their care.

*

Brand concluded that "shared pain is central to what it means to be a human being," but we are a society that values the anesthetic over pain. We hide our prisons, our sick, our mad, and our poor; we expend colossal resources to live in padded, temperature-controlled environments that make few demands on our bodies or our minds. We come up with elaborate means of not knowing about the suffering of others and of blaming them when we do.

Choosing not to feel pain is choosing a sort of death, a withering away of the expansive self. When Robert Jay Lifton went to investigate the psychology of survivors of the atomic bombings of Hiroshima and Nagasaki, he coined the term "psychic numbing" to describe the survival strategy of dissociation and apathy—"a diminished capacity or inclination to feel." In such extreme circumstances it was necessary or at least understandable, but even there Lifton called it "dehumanization" and cautioned that it "comes to resemble what has been called 'miscarried repair.'" He compared it with immune disorders that begin by eradicating outside elements and then turn on the body itself. Decades later, when he looked at the numbing of those who used the atomic bombs and those of us in whose name they were used, he reverted again to medical metaphor: the cordon sanitaire, "a barrier designed to prevent the spread of a threatening disease, the 'illness' we block off in this case being what we did in Hiroshima."

We think of kindness as an emotional quality, but it's also an act of imagination, of extending yourself beyond yourself, of feeling what you do not feel innately by invoking it. This is pretty instinctive when we watch a child skin her knee, perhaps less instinctive when we read statistics on Haiti or Syria and have to translate them into feeling. You could call this feeling love, for we suffer with and for those we love, and we seek to protect them from suffering.

Imagination enlarges us—as though our nervous systems could be made vast and at home in the world, if not at ease with its cruelties and losses. Comfort is dangerous. You can be overwhelmed by suffering, as relief workers sometimes are, and your ability to imagine and engage is finite—as anyone who deletes all those e-mails urging us to act for prisoners or polar bears or disaster victims knows.

Daniel Ellsberg decided that stopping the war in Vietnam, whose hundreds of thousands of deaths became real to him after he saw some of them firsthand, was more important than his own well-being, and so he risked his freedom and gave up his government career to expose the truth. In a 2009 documentary film about his actions as a whistleblower, Ellsberg said that after he released the Pentagon Papers in 1971, his coworkers at the RAND Corporation treated him "like a leper." In his memoir, *Secrets,* Ellsberg recounts one of his colleagues' telling him that he agreed with the leak in principle but not in practice. Publicly supporting Ellsberg, he said, would have forced him to "renege on . . . my commitment to send my son to Groton." The colleague had weighed many sons' violent deaths against his son's prep-school education and his own career, and the latter remained more real and more compelling than the former.

Up close, aggressive measures are required to be impervious to suffering; you have to convince yourself that people deserve what they're getting, that their suffering has nothing to do with you. Our capacity for empathy is why the reality of war is usually kept from us or delivered in measured, manipulative doses—our wounded, perhaps, but not theirs, or those of our wounded who make for uplifting stories, but not those who are severely mutilated. And we face choices about how we live, because we are implicated.

Sometimes I'm tempted to bemoan the abundance of suffering in the world and leave it at that. But there is also a constant supply of kindness in circulation—or so I came to believe after spending time in January 2010 at the Hansen's disease clinic in Baton Rouge, where Eddie Bacon went to be treated. The clinic occupies the fluorescent-lit, fake-wood-paneled second floor of a medical building, a labyrinth of rooms that wrap all around the elevator bank and are abuzz with activity.

Since treatment of the bacterium is now simple, the bulk of the clinicians' work involves teaching the patients how to take care of themselves after they are cured. Pam Bartlett, a social worker, meets with those just beginning therapy—she's had to talk several out of suicide—and then keeps track of them throughout their treatment. There's a surgeon on staff, Ronnie Mathews, a pupil and disciple of Paul Brand. Many patients get fitted for special shoes, built or adapted to protect their feet from their own insensibility; a shy young man named Jonathan Starks was in the shoe-making room on the day I visited, preparing to take over a trade his grandfather had started. All treatment is free.

I also met Barbara Stryjewska, the clinic's chief medical officer, a merry redhead in high heels who tends to touch and hug and tease her patients. She showed me an astounding photo album of them and their afflictions before and after treatment, along with images of an amputation. (She seemed not to consider at all how this might look to an outsider.) When I met her she was visiting with the Thompsons, a Cajun couple who told me how frightened they had been when Vicky Thompson was diagnosed with a disease they knew only from the Bible. They ended up demonstrating some swamp-pop dance steps for me in the office of the staff epidemiologist. Stryjewska told me that forming long-term relationships with patients was the only way to treat them effectively. Part of her job is to figure out what habits contribute to poor health —in the case of Vicky Thompson, it was six-hour dance sessions that were hard on her feet.

It's what one might call a new form of holistic medicine: they teach patients how to take care of wounds, how to look for injuries, how to avoid burning their hands when cooking. It's an empathy factory, and it bustles. While there I met an octogenarian doing physical therapy to strengthen a numbed leg, a shy teenager from Micronesia who looked as if he had nothing more than a bad case of acne, a sturdy middle-aged Latino guy from Los Angeles with a persistent foot sore who reclined on his hospital bed like a Roman emperor. Staff members stopped by to lay hands on his bare shin and chat or to look at his feet—one of his big toes was swollen to the size of a potato. Ronnie Mathews came by on a sort of improvised walker—he's disabled himself—and whittled at the toe a little, casually, like a man giving a pedicure with a scalpel. (An advantage of working on leprosy patients is that you can often skip the anesthesia.)

I met Dane Hupp, a physical therapist who told me he sometimes chewed out the patients because "I treat them like my family." He was worried about a patient who, if he didn't get over some bad habits, could lose his leg—though, he told me, they lost a lot more legs at the diabetes clinic where he used to work.

Irma Guerra, who runs the outpatient clinics from an office across the parking lot, told me she still thinks about Eddie Bacon. "We learn so much from people like him because they suffer. You can see how beautiful his personality is, what courage it takes to be like that. The people I've met I'll never forget." Mathews told me he thought about Eddie, too. "I wake up some nights and wonder how he's doing."

Captain John Figarola, of the U.S. Public Health Service, is head of rehabilitation at Baton Rouge. He is a tall, earnest New Orleanian with brown eyes and a beige uniform with small metal insignias on the collar tips. He talked to me at one end of a long table while Mathews delicately pared the infected finger of a nervous young man from Latin America. The surgeon was trying to cut away all the dead tissue on the finger before a bone infection further shortened the digit. "My friend," he said, "I'm afraid if I send you home, the tip of the bone will get reinfected."

Figarola told me that he loved his job because he and his staff are allowed to give care without measure—they can act on empathy. As Mathews worked, a little crescent of fingertip landed on the table between Figarola and me. "I work in utopia," he told me, and threw the scrap away with a bit of paper tissue.

DAVID TREUER

Trapline

FROM *Orion*

BEAVERS ARE, as far as animals go, odd contraptions. The largest
rodent in North America, the beaver has webbed feet, a scaly tail,
and two front teeth with orange enamel on the front and dentin
in the back so that as they wear down they self-sharpen. They are
powerful swimmers, chewers of trees, and builders of dams—some
of which have been known to stretch for hundreds of feet. They
were once trapped in unsustainable numbers for their fur, their
fat, and their scent glands, which produce a secretion that has
been used for medicinal purposes since antiquity. Pliny the Elder
maintained that the smell of the glands was so powerful, much
like smelling salts, as to cause a woman to miscarry, although it was
later diluted in alcohol for use as a musky addition to perfumes.

When beavers were plentiful during the early days of the fur
trade, my tribe, the Ojibwe, enjoyed an incredible quality of life.
While other tribes were being wiped out or displaced, our birth
rates were up and our land base was increased by a factor of twenty.
In 1700 England exported roughly seventy thousand beaver-felt
hats (beaver skins were dehaired, and only the hair was used in
hat making). In 1770 the number of exported hats had risen to 21
million. But the demand (and supply) of beaver couldn't last.

Some estimates place the number of beavers in North America
at over 60 million at the time of contact. By 1800 they were all but
extinct east of the Mississippi. We Ojibwe shared much the same
fate, pushed west, reduced in numbers, eking out an existence in
the swamps and lowlands of the American interior. Though the
tribe was once as defined by trapping beaver as the Aztecs were

defined by gold or the Sioux by the buffalo hunt, by the twentieth century only the idea remained. The furs and the knowledge necessary to harvest them were fading.

As an Ojibwe child from Leech Lake Reservation in northern Minnesota, I grew up around hunters. But aside from our mother teaching us to snare rabbits, we didn't trap animals, as "bush Indians" did. Instead we harvested wild rice in the fall, made maple sugar in the spring, and shopped for food like everyone else. My father (a Jew and a Holocaust survivor) and my mother (an Ojibwe Indian and tribal court judge) put as much emphasis on homework as they did on living off the land. None of the other kids from Leech Lake that I knew had grown up trapping or living off the land either. But this was still held up as the only truly Ojibwe way of life, and as I grew older I longed to commit myself to the bush. So when Dan Jones, an Ojibwe friend from across the border, offered to teach me to trap beaver on his trapline in northwestern Ontario, I said yes.

Lewis Henry Morgan in 1868 wrote, "The life of the trapper, although one of hardship and privation, is full of adventure. They lead, to a greater or less extent, a life of solitude in the trackless forests, encountering dangers of every kind, enduring fatigue and hunger, and experiencing, in return, the pleasures, such as they are, afforded by the hunt."

Perhaps because I had never been trapping, I was sure that every word Morgan, a gentleman ethnologist and fan of the beaver, wrote was true; at least I wanted them to be true. I wanted the special knowledge that all trappers seem to possess. I wanted their forearms and their expertise with a knife. I wanted to be a part of that brotherhood of whom it was said in bars, and around kitchen tables, and over the open tailgates of dented pickups, "Oh, him? He's a real bushman. No one knows the bush like he does."

I drove to Dan's reserve in northwestern Ontario after Christmas in 1996. The name of the reserve in English is somewhat odd: Redgut Bay (named after a former chief of the band). The Ojibwe name is much longer: Nigigoonsimini-kaaning (The Place of Abundant Little Otter Berries). I have never been able to find a "little otter berry," nor have I found someone who has found one, so part of me wonders if "little otter berry" is a way of saying "otter shit." Which just goes to show that if you scratch the surface of romance you'll find slapstick, and if you scrape off the slapstick you might find wonder, because after all, places (like animals) don't always give up their secrets.

All of this—romance, slapstick, wonder—were mingled in Dan. About five foot eight and more than two hundred pounds, Dan looks the way a traditional Ojibwe man should look: stocky, strong, black hair, dark skin. He's also—and I've tested this—pretty close to imperturbable. Once, when we were checking traps together on a beaver house, the ice broke underneath him and he fell into the freezing water. All he said after he got out was "oops." He is indifferent as far as money is concerned. I've never known a stronger paddler. When you see him filleting fish or skinning animals, you think to yourself that he was born with a knife in his hands, yet when he sleeps he needs no less than three pillows to be comfortable. His jokes are terrible. He thinks it great sport to tease people in uncomfortable ways in public. He remains one of my very best friends and always will.

Dan mostly traps for two animals—beaver and pine marten (like a Canadian sable)—although he grew up trapping and snaring just about anything that moved. His mother once held the world record for beaver skinning. He was raised on the trapline, moving from the village out to the line in the winter, back to the village in the summer, to rice camp in the fall, and back to the trapline. His first memory is of lying in a rabbit-skin sleeping bag, watching the jagged outline of spruce against the sky.

We drove to the supermarket and loaded up on what Dan referred to as "trapping food"—cigarettes, bacon, eggs, butter, bread, Chips Ahoy!, Diet Pepsi, canned potatoes, oatmeal, and pork chops. His wife drove us out of town on a double-track path through stands of jack pine and over frozen creeks. We followed it for 12 miles and stopped. We offloaded the snowmobile, attached the sled, and filled it with food and clothes, then drove off the road toward Moose Bay—a clear, clean fingerlet of water stretching to the northernmost arm of Rainy Lake. An hour later we arrived at the cabin on a small bay surrounded by balsam and poplar.

Rainy Lake is in the Canadian Shield. There is water everywhere —pond after pond, river after river, lake after lake. According to geologists, the water is still learning where to go, channels and streams hardly set. This land of old rock and new water forms the base of the boreal forest—the largest unbroken forest in the world

—and is the world's largest terrestrial biome. The land is studded with pine, fir, and spruce. In summer, it is almost impassable; in winter, if you step off your trail, 100 yards might as well be a mile through the deep snow.

There might be something about the Great Plains—the openness, the sense of scale—that is good for stories and epic struggles. Not so the boreal forest. Horizons are hard to come by. The sky is a fractured thing. There are precious few vistas. Instead you are enclosed, hemmed in, covered over. It is a good place for secrets and secret knowledge, for conspiracies and hauntings.

The cabin was less beautiful than the country around it and, as regulated by law, rather small—ours was 16 feet by 20 feet. Only trappers who buy the trapping rights to an area are allowed to build one. This one hadn't seen a human being since the previous year. If what happens on the trapline stays on the trapline, then it's equally true that what goes into the cabin stays in the cabin. Old cupboards were shoved in the corner between two nonworking gas ranges. A table, three beds (one a stowaway bed like you find in hotels), clothes, a wood stove, a box of beaver traps, and three or four bags of garbage completed the cabin. Under one of the beds I found a stack of *Playboy* and *Penthouse* magazines; a centerfold was tacked to the door. The whole place was overrun by mice. That night we cut firewood, got the cabin thawed out—it was minus 10 degrees Fahrenheit—cooked some pork chops, and chopped a hole in the lake to get water. Then I learned the first thing about trapping: how to play cribbage.

Trapping beavers is largely a matter of finding where they live. Before ice-up, either by walking along the shore or by canoe, the trapper will locate the beaver house, beaver dams, food stash, and channels (worn paths near convenient food sources) and place drowning sets at these areas using either 220 Conibears or 330 Conibears. This trap was invented about fifty years ago by Frank Ralph Conibear, an Anglo-Canadian trapper. It was a revolution in traps and trapping, making it much more certain and productive. Conibears, or body-grip traps, are two steel squares attached to one another by a hinge. Steel springs keep the jaws open. To set the trap, the springs are compressed and the steel squares of the trap are held together by a catch from which dangles a trigger. If a beaver swims through and touches the trigger, the catch moves up, and the springs, under tension, slam the steel squares open across the beaver's chest, neck, or head. Death is quick. Until the advent of body-grip traps, trappers relied on wire snares or, more often, leg-hold traps set in channels and at the entrances to a beaver lodge. These would close on the beaver's foot and drown it, or, often enough, the beaver would lose a leg in the trap.

Before steel or iron, trapping was another matter entirely. To trap beaver without the use of steel meant tearing open beaver houses to catch the beaver in its den, which is a hollow chamber above the waterline, or isolating the channels and runs under the ice. One would break open the house and wait at each and every channel for the beaver to surface, then kill it—with arrows, guns, or clubs—when it emerged. This was time-consuming and brutally hard work, and it took a lot of bodies—two people tearing up the house and four or five waiting at the channels.

But Dan and I had everything we needed, and we relied on steel body-grip traps exclusively. We would be trapping the entrances to the beaver houses on the string of ponds that, like terraces, are stacked one on top of the other all the way from the big lake deep into the woods, almost to where we were dropped off. We stopped at the first beaver house. Dan showed me how to tell if the house is occupied or "dead": look for a cone of crystallized vapor on top of the house that looks like a nipple or wick. This is a sure sign that the house is live; the beavers' warm breath travels through the frozen slurry of mud and crisscross of sticks on top of the house. When it meets the subzero air, it freezes, creating the nipple. In the first pond, one house was live and the other dead. Once we located the live house, we got off the snowmobile, and with Dan in the lead, we tapped around the house with the point of the chisel. The ice is usually thinner over the entrances, worn by the passing in and out of the beavers' bodies. The chisel broke through, we cleared the ice and sticks away, and using a long bent pole, found the beaver run. We set our traps. Then on to the next, and the next, and the next. We put thirteen traps in the water that first day.

Perhaps this is the strangest thing about modern trapping among the Ojibwe: it is an age-old cultural practice, as ingrained, as natural, as everything else about our culture. The snare, the trap, the trail, the lure, the catch—these are the metaphors by which we make our meanings. Yet it has been many hundreds of years since we have even so much as worn a fur, except as decoration or for ceremonial purposes. As soon as we could, we traded pelts for guns, axes, kettles, wire, and cloth. Furs are fine, I guess (I have, for sentimental reasons, a beaver-skin cap and moose-hide gloves). But cloth—wool especially, but also cotton—lasts longer, is easier to clean, can be sewn into many more things, and holds up longer. I only know of one Indian who wears furs—Jim LaFriniere of White Earth Reservation has a muskrat-hide jacket. I don't think he wears it because he's Indian; I think he wears it because he is, at White Earth, a BIG MAN. Modern synthetics are even better than trade cloth. There is nothing quite like chopping a hole in the ice with an ax and getting covered in dirty slush, only to have it bead and then freeze on my Gore-Tex parka. As for gloves, moose skin gets slick fast when the water freezes, and the ax goes flying out of your hands. But waterproof gloves with rubberized palms—they make all the difference. All the furs we catch—with a few held back for ceremonial use—will go on the market. In 1996 beaver were fetching, as I remember, between thirty-five and fifty Canadian dollars per hide. Occasionally we ate the meat. Mostly we ate pork chops.

The next day we had three beaver in our traps. Not bad for one night's sets. We checked the traps all that morning, napped, and then spent the evening skinning. Or, rather, Dan showed me how to do it on one beaver, and I spent the rest of the night working on the other two (since they were caught in sets I had made). As with any kind of hunting or trapping, the killing is the easy part; skinning and butchering are where the work is. Beaver have incredibly thick hides, which, when rough-skinned (taken off the body but not cleaned), are thick with fat, especially around the tail, that I can only describe as blubber.

I learned quickly to rough-skin a beaver in ten minutes or less. Fine-skinning, or fleshing, is much harder and more important —leave too much fat on, and the hide won't dry right; slice or put holes in the hide, and that lowers its value. There are many methods for fleshing. Dan pinches the hide between his thumb and middle finger, and using his index finger to provide tension, he takes long, smooth strokes with his knife; the fat peels away smoothly and cleanly. It took me hundreds of hours of practice to approximate his skill.

As I struggle to separate the fat from the skin using a fillet knife,

Dan, smoking a Player's Light and drinking Diet Pepsi, tells me that all the furs are taken to the fur buyer, graded, and bought, and then the fur buyer takes them to a fur auction where lots of furs are bid on, purchased, and then sewn or made into something. Since the British have stopped wearing funny-looking beaver hats, I'm not sure who's buying them. Dan says the biggest buyers are the Greeks, Russians, and Chinese.

It was uncomfortable, to a degree, to see the beavers undergo the transformation from beautiful animal to a skin worth x dollars. Hunting is, to many people, more palatable, I suppose. Eating an animal you killed seems more just. On the other hand, maybe it only seems that way. It is largely a myth that Indian people were somehow natural conservationists who used all the parts of the animal. We were as wasteful as every other people living on the move without electricity or refrigeration: we ate what we could, dried what we could, and left the rest for the wolves. And trading an animal's life for the resources that you need is not, as far as Dan is concerned, a bad trade. His response is that if our ancestors had had the same ethical concerns, we wouldn't be here today. (By comparison, my father—of European stock—had no use for trapping. But the "old ways" of doing things were one of the things my father admired, perhaps romanticized, about the Indians he befriended when he moved to Leech Lake in the 1950s.)

Dan was clearly less worried about it than I was. And why should he be worried? Why, just by virtue of being American Indian, should he live out ideals (about the sanctity of life, about the equality of animals) that have largely been foisted on him by James Fenimore Cooper and Rousseau and every other conscientious outsider? Dan is an Indian who loves to trap, who loves the animals he traps, enjoys the process of handling them, and who, at the end of all that, loves to golf and needs new clubs.

After the hides are fleshed, they are nailed onto boards on which a series of concentric ovals have been drawn. The size of the beaver determines which oval you use. Stretching beaver is its own art. Too loose and you cheat yourself of profits because the beaver dries to a smaller size than it might have. Too tight and the number of hairs per inch is reduced, your furs are graded down, and you lose money. We leave the beaver carcasses on the ice for the eagles and the wolves—there's not a trace left come morning. The fat and muscle skinned off the hides are chopped into baseball-sized chunks as bait for marten and fisher (a sort of cross between a marten and a weasel but much larger).

New Year's passes, and we listen to the country countdown on the radio, play cribbage, skin and stretch beaver. Every day is blessedly the same. There are no other people. Nothing moves. Occasionally we hear a plane far overhead or a snowmobile in the distance. I hear wolves at night. We have caught sixteen beaver and a few marten by the first week of the new year.

And then Dan says, "Think you can handle this on your own now? I've got to bring our furs back to the reservation, and then I have to go to work. I showed you how to do it beginning to end. No problem, right?" He leaves the next day, and I will be, for the next two weeks, on my own. Since he is taking the snowmobile with him, I will also be on foot.

The days blended into one another. I left the cabin after first light carrying a shotgun, an ax, a pack with lunch, and coffee. I was finally getting the romance I thought I wanted. Our line of traps was 17 kilometers long, and I walked it every day. I checked and reset the traps, skinned the beaver on the ice and put the hides in my pack, caught the occasional rabbit in a snare, shot the occasional partridge. I would return to the cabin after crossing a wide bay of Rainy Lake and sleep for a few hours before fleshing and stretching. After that I read, wrote, and went to bed. I bathed by heating up water and pouring it into a five-gallon plastic bucket. I saw a lot of trees. I saw a lot of snow. I caught a lot of beaver, and I skinned a lot of them. I read and reread *Confessions of Felix Krull,* and when I ran out of cigarettes I began ripping out pages from the back to roll tobacco and read the *Playboy*s instead. I read Tim O'Brien and T. C. Boyle and marveled at the odds of two of my favorite stories —"On the Rainy River" and "King Bee"—existing in this trapping cabin far from any other kind of print. I began to dislike airbrushing for the same reason that long ago I really liked airbrushing.

During the nights I listened to the ice booming on the lake. During the day, when I went to fetch water from the hole in the lake, I began to see how subtle changes in temperature and wind affected the thickness of the ice. I heard a lot of wind and came to like the difference between wind through spruce, wind through balsam, wind through bare poplar, wind through red oak, wind through marsh grass, and wind through dead cattails. I saw what wind and sun did to old moose tracks and deer tracks and squirrel tracks and rabbit tracks and fisher tracks and marten tracks and fox tracks. I once crossed a pond to check my traps, and when I crossed back ten minutes later, seven sets of timber wolf tracks had crossed over mine. I learned that, despite everything, I wasn't very comfortable with the idea that there were so many timber wolves so close to me. At night, when I fleshed and stretched the hides, I listened to country music. I memorized "Strawberry Wine" by Deana Carter, "Is That a Tear" by Tracy Lawrence, and "Little Bitty" by Alan Jackson. I liked "Little Bitty" least of all, but I found myself singing it more often than any other.

I learned I liked quite a bit the mediciney smell of beaver fat. I learned that each and every animal I killed and then skinned was more or less perfcct. I learned that each and every animal had been designed to live a certain way and had acted according to that design. I learned that walking upward of 17 kilometers a day, chopping through inches of ice, cutting firewood, and hauling water on a diet of pork chops and oatmeal gets you in very good shape. And then, one day, I learned that steel was a pretty amazing thing and that without it very little of the bounty around me would be mine.

I had been chopping through the ice to check one of my beaver sets. Each night a couple of inches of ice formed over the hole, and so every morning I had to remove that ice. It was soon mounded all around the hole, and the hole itself was like a funnel. I had finished chopping and had scooped out the ice and slush and placed the ax behind me, and before I knew it the ax slid down the funnel and disappeared into the water. It was the only ax I had, and without it I wouldn't be able to check any of the other traps. Without it I couldn't split any wood for the stove, and dry wood was scarce near the cabin. Without the ax, I would have little to do and our fur count would plateau and I would be reduced to eating out of cans. In a flash, I came to appreciate my tribe's age-old hunger for metal and later plastic. One of the great criticisms of my tribe's behavior in the eighteenth and nineteenth centuries was our so-called dependence on trade goods. But try living without metal knives, or axes, or even a pot to cook in.

I knew I had to do something about the ax, and there was only one thing to do: I stripped down, set my clothes on the ice, and lowered myself into the hole. The water was 34 degrees, and it hurt everywhere at once. The good news was that the water wasn't any deeper than my armpits. The bad news was that I had only a few seconds during which I'd have feeling in my toes. I found the handle with my left foot, took a breath, ducked under the surface of the water, grabbed it, and got out as quickly as I could. It took me an hour of fast walking to get warm again.

Dan came back after two weeks and brought with him more coffee, more cigarettes, more food, and the feeling that instead of romance I had gotten intimacy. With him, to be sure. But also with the animals under the knife and the place itself. After a while I had to try very hard to locate the danger and excitement that writers like Morgan and many others attach to the bloody business of professional trapping. For Dan, growing up on the trapline and then returning to it after high school had been the most peaceful times of his life. Indian boarding school, life in the mainstream, these had been bloody and hard. Trapping, by comparison, was guided by rhythms and activities that were, in themselves, small, finite, measurable, and, paradoxically, eternal—a quiet, steady kind of work that was reminiscent of a life outside of time. For me, after a while the thrill of trapping gave way to a deeper satisfaction much harder to name and much more profound than romance and danger.

This became the rhythm of the rest of the winter: two weeks alone, Dan for an extended weekend, repeat. It was part of the most profound years of my life, at the end of which there were many things I could do that few others could, and many things I could do that I never imagined I could. And none of what I had learned really mattered in the larger world. I was pleased to discover that trapping rewards a mind that is organized, creative, and neurotically interested in details. Which is to say: trapping is an activity made for a mind just like mine. I became a trapper.

For the next five years, I spent a few weeks every year trapping with Dan and a month or two trapping on my own back home on the Leech Lake Reservation. In 2002, after having trapped beaver, mink, marten, fisher, and otter, my brother Micah and I decided to expand our trapping techniques to include more snaring. We hoped to snare a fox. We talked to as many trappers as possible, read books, went to trapping forums online, and after buying and treating (boiling, dying, and waxing) our snares, we were ready to begin.

We hung our first snares at the beginning of the holidays in December. It was warm and there wasn't much snow and we didn't know what we were doing. We set our snares too high or too low; we set them on rises and humps so they were too clearly silhouetted on the trails. We set a snare in an area that we thought was a fox run and came back the next day to find a porcupine caught by the neck and foreleg. He squirmed in a kind of slow agony. We used natural funnels—places where the game trails narrowed to squeeze through dense brush or swamps or between deadfalls —but we didn't trust the snares themselves and blocked up the trail with sticks and branches so that the fox would have nowhere to go but into our snares. It must not have looked right to them: no fox came near our sets. We did almost everything wrong. Every morning we got up, excited at the prospect of fox after fox dead in our snares. Day after day our snares hung empty. Christmas came. It went.

On the twenty-sixth of December our mother called all of us over to her house—her partner, Ron; my siblings; and our spouses. She had news, she said: she hadn't been feeling well for some time. She had been coughing a lot. Her ribs on her right side hurt her constantly. She was tired and had lost a lot of weight. She'd gone to the doctor, and they had taken X-rays and made scans and had detected a large lump, a tumor, in the lower lobe of her right lung. The tumor had grown so large as to push past and envelop her ribs. These had become brittle and had, at some point, broken —the source of the pain. They had taken a sample to be biopsied and she would, she said, know more soon. She had been a steady smoker for over forty years. It was, in all likelihood, lung cancer.

I can't remember how we reacted. Some cried, I'm sure. Some didn't. Some started strategizing—as though the cancer were an enemy we could fight. I think this was probably my response. It would be a week before we got the results of the biopsy. In the meantime we carried on. We got groceries. We went to the bank. We watched movies. We argued. We did everything we could. We did nothing at all.

When Micah and I went to check our snares, I noticed that the fox we had been trying to snare had begun using the ruts my truck tires left in the fields of big bluestem we crossed. So, on a whim, we cut down a small jack pine, dragged it close to the tire tracks, and wired a snare to it.

On New Year's Eve my mother gathered us together again and told us that she had lung cancer. The doctors planned to operate within the week. I don't remember much of that time. I don't remember living in any conscious way—that is, making decisions or acting purposefully—but I must have. I do remember thinking a lot about snaring.

Snares are elegant tools—there is something beautiful about a snare, whereas there is little that is beautiful about a metal trap. Metal traps, no matter what kind, are nasty, brutish things. A snare is so simple: a piece of wire formed into a loop. One end is anchored to the ground or to a drag stick; the other ends in a lock through which the wire passes. When an animal walks through a snare, the lock slides down the wire and the snare tightens around the animal's neck. The animal struggles, and the snare gets tighter and tighter, until the animal can no longer breathe and it dies. With a piece of wire and not much more than that, a man can survive for a long time.

As I contemplated my mother's operation, it seemed to me that seen in a certain way, snares, unlike traps, don't actually kill the animal. The animal's habits are what kill it. Same with my mother. Every animal has its habits—where it walks, where it hunts, where it dens, where it mates—and snares more than any other kind of trap take advantage of those habits. Trapping beavers at their houses or dams or channels is a matter of taking advantage of geography—if you put a trap in a doorway, the beaver will have to go in or out sometime. Fisher and marten traps are baited and as such use hunger against the animal (a weasel must eat two times its body weight a day just to stay alive). But snares use an animal's habits. Instead of using lures or attractants or bait, instead of trapping the entrances of a den or digging out the den or burrow, you set a snare where an animal goes, and a good snare set works because it is unobtrusive. And there they hang—a nice clean loop, no mechanical springs or parts, nothing but gravity and the animal's own struggles to help with the kill—and could hang forever, waiting. It is tempting to think of snaring (and trapping) metaphorically—we already speak of things like "snares of love," for example—but the real beauty of it is literal: this wire, on this trail, will choke to death a fox whose life is not that of all foxes, but was his and his alone.

Four days after my mother received her diagnosis, we went, as we did every day, to check our snares. We drove out into the fields and checked our line along the old fencerows and among the jack pine and down near the slough. Nothing. We drove back out into the fields and Micah yelled, "Stop, stop!" I stood on the brakes, and above the dead brown grass we saw a fox jumping and twisting. He would disappear into the grass and then jump in the air and fall back down. We bailed out of the truck and ran toward it. He was caught in the snare we'd set in the open field on the tire track. It was a clean catch. Tight around his neck. But he must have gotten into the snare just a short while ago. The stick to which the snare was wired was too big for the fox to drag very far.

A few days later my mother went into surgery; they resected her ribs and removed the lower two of the three lobes of her right lung. All of us were waiting for her in intensive care when they wheeled her bed in. She was still unconscious and on a respirator. A long translucent tube snaked from a hole in her side down to a bag filled with blood and a slimy yellowish fluid. She was gray and ashen, and though she wasn't awake and couldn't have spoken if she were, the way the tube went down her throat distorted her face. She looked like she was screaming. But her body was limp, her eyes shut. The only sound was that of the respirator and the squeak of our chairs. I wanted her to live. I wanted it more than I had reason to expect she would. I closed my eyes and tried to think of something else—of something other than her pain, and whatever the future might be, something other than our collective hopelessness. I could think of nothing.

The fox we'd snared had also wanted to live. That, after all, was his purpose. He'd wanted it so much that when he felt the snare tightening and he couldn't breathe, he tried to run away, to get his body far from the snare and the log to which it was attached. He jumped again and again, and it was something both strange and beautiful. He lifted clear into the air—a bright flash of red against the sky—and then disappeared below the grass, which was about 3 feet high. First his nose, then his body, and then his black-and-white-tipped tail cleared the grass and was jerked back down by the weight of the log. He jumped and jumped and jumped again. All his traits and everything he had learned, the land itself and what it offered him, forced him to choose this path, on that patch of land on that day, and it was killing him. His instincts were killing him, but it was his instinct to live, too.

Finally we drew close enough to knock him on the head with the ax handle and down he went. I felt the quickness of his breath as I knelt on him with one knee. With one hand on his head and the other on his chest, I felt his heart and the life in it. Who knew a heart could beat that fast? I felt, too, in those beats and under that fur, in that quick, elegant body, how much it strained toward life, how much it jumped for it. Everything in that animal's body was bent on it. It wanted to live. And we, too, gathered around our mother, wanted to live. And she wanted to live. We and all the others and everyone—regardless of the lives we'd led, and more than anything else, and beyond the agonies and dangers that attend every act and action of ours in this life, we all wanted to live. And that desire, if not the result, is something to think about.

E. O. WILSON

The Rebirth of Gorongosa

FROM *National Geographic*

IN THE SUMMER MONSOON SEASON of late November to mid-March, the rain clouds ride the trade winds of the Indian Ocean west into Mozambique. Crossing the coast, they refresh the *miombo* woodlands of the Cheringoma Plateau, then the savanna and floodplain grasslands of the Great Rift Valley. Finally they run aground on the slopes of Mount Gorongosa, where they release great torrents of rain, like a benediction.

The Gorongosa massif, which reaches a height of 6,112 feet, captures more than 6 feet of rainfall a year. That is enough to support a lush rainforest on the summit—and, to the east, in the Rift Valley, a park that was once one of the richest wildlife refuges in the world. Before Mozambique's civil war ravaged it, Gorongosa National Park was roamed by elephants, African buffalo, hippopotamuses, lions, warthogs, and more than a dozen species of antelope. Now some of those animals are coming back, thanks largely to Greg Carr, an American businessman and philanthropist who is leading a project to restore Gorongosa. In 2010 the park marked a milestone: Mozambique's government fixed an error made at its creation, expanding its boundaries to include Mount Gorongosa, source of its life-sustaining rivers.

In the summer of 2011, I went to Gorongosa to support Carr's efforts and also to work on my new digital textbook for high school biology. The park is an excellent place to convey the high stakes and the excitement of doing wildlife biology today. The summit rainforest on Mount Gorongosa, about 29 square miles in extent, is an ecological island in a sea of savanna and grassland. It is hard to get to, and so it has remained largely unexplored by biologists. Ants, my own specialty, were entirely a blank on the map when I arrived. For a naturalist there is no more powerful magnet than an unexplored island. When I visited Mount Gorongosa, on my first trip to Africa, I felt highly charged with the prospect of surprise and discovery.

During my stay at the park my assistant was Tonga Torcida, a young man born on Mount Gorongosa. He was one of the first of his village to graduate from high school, not a mean feat, since schooling past the seventh grade requires tuition and a uniform that few local families can afford. While we were together, Torcida learned that he would receive a scholarship to attend a Tanzanian college. Speaking four languages and working off his intimate knowledge of Gorongosa, he plans to be a wildlife biologist.

Torcida told me a creation story of his people and why they consider Mount Gorongosa sacred. In early times, he said, God lived with his people on the mountain. Humans were giants then and not afraid to ask God for special favors. In a drought they would say, Bring us water. The Creator, growing tired of their constant importuning, moved his residence up to heaven. Still the giant people persisted, reaching up from the mountain. At last, to put them in their place, God decided to make them small. Thereafter life became a great deal more difficult—and so it has been to this day. I told Torcida that this folklore and the moral lesson embedded in it sounded very much like parts of the Old Testament.

Gorongosa certainly suffered a precipitous fall from grace. Three years after Mozambique won its independence from Portugal in 1975, a civil war broke out and raged for seventeen years. The park, which had been established by the colonial government in 1960, became a battleground. Its headquarters and tourist facilities were destroyed. Roving soldiers, hungry for food as well as for ivory they could trade for weapons in South Africa, killed many of the large animals. After peace accords were signed but before order could be restored at Gorongosa, commercial poachers killed an even larger number of animals, peddling the meat at nearby markets. In the end, nearly all the big-game species were gone or nearly gone. Only the crocodiles, quick to slide down muddy banks into the safety of the rivers, escaped with little harm.

The clearance of big game had important environmental consequences. Where zebra herds no longer grazed, grass and woody shrubs thickened, and lightning-strike wildfires became more threatening. With no elephants knocking over trees to feed on the branches, some forests increased in density. With the scat and carcasses of big game severely reduced, the population of some scavengers fell sharply.

Yet the ecological base of vegetation and small animals, including the myriad species of insects and other invertebrates, remained largely intact. Gorongosa Park contains a great variety of habitats —besides the valley grasslands and the mountain's several vegetation zones, it includes forested plateaus and limestone gorges— and even today it supports tremendous biodiversity. In the whole of the park, 398 bird species (of which about 250 are residents), 122 mammals, 34 reptiles, and 43 amphibians have been found. Probably tens of thousands of species of insects, arachnids, and other invertebrates await discovery.

For a decade following the end of the civil war, while a new, democratic Mozambique established itself, Gorongosa remained in ruins. Meanwhile, Greg Carr had gotten interested in the country and was looking for a way to help; after making his fortune in voice mail and Internet services, he had turned to philanthropy. In 2004 the government of Mozambique and Carr agreed that he would help plan the park's restoration. Carr has since done much more: he has undertaken to restore Gorongosa himself, largely at his own expense, and has made it his full-time occupation. Mozambique's Ministry of Tourism has entered into a long-term partnership with him to manage and develop the park.

Today, after less than a decade, Gorongosa is well on its way to recovery. Large animals, including African buffalo and elephants, have been imported from nearby South Africa and are multiplying rapidly. Eland and zebra are next. Though still well below their prewar maximum, herds of grazers and browsers swarm once more across the savanna and grassland. Ecological balance is returning with the megafauna, and so are visitors from Europe and North America. Excellent facilities have been built at the central Chitengo Camp and at explorer camps in the interior. At Chitengo a bullet-pocked concrete slab has been preserved as a war memorial.

The accomplishments of Greg Carr's team and of the people

of Mozambique are impressive. But restoring a damaged park is much harder than creating a new one, and Gorongosa—especially its mountain—is far from being out of danger. During the civil war, as marauding soldiers invaded the mountain, subsistence farmers began to clear little plots up the slope. The taboo of the sacred mountain was largely forgotten. In time the farmers reached the summit rainforest and began to fell the tall trees and convert the moist, fertile ground into corn and potato fields. In the past decade the area of original rainforest has been reduced by more than a third.

The retreat of the forest already means that fewer species of plants and animals, some likely endemic, can survive. The complete removal of the forest, which at the current rate of destruction might easily occur within ten years, would be catastrophic for the entire park. The mountain's ability to capture, hold, and gradually release monsoon rainwater would be gone. The water would then run off quickly, and the moisture supplied to the rest of the park would be rendered seasonal instead of year-round. In the face of the new aridity, life in and around the park would be less sustainable for both wildlife and people.

Now that the mountain is part of it, the park has the authority to secure the forest perimeter. The forest won't be truly secure, though, until those who are destroying it are given alternatives. Tourism is part of Carr's answer, but he has also hired teams to create numerous nurseries to grow seedlings of the rainforest trees and begin the decades-long, perhaps centuries-long process of returning the forest to its original area. The park is creating schools and health clinics for local people at the base of the mountain, below the rainforest. Finally, a center for scientific research and education is planned for the Chitengo Camp. The emphasis will be on the environment inside the park and on the preservation of its biodiversity.

To sample the current biodiversity on Mount Gorongosa, Greg Carr and I decided to hold a "bioblitz" there and to engage the community living on its lower slopes. We asked Tonga Torcida to help organize the event and to recruit local children as our helpers. Bioblitzes are counts of species found and identified in a restricted area over a fixed period of time, usually twenty-four hours. They follow simple rules: participants search within a set radius around a focal point, assisted by local naturalists who are familiar with one or more groups of organisms and can identify the species discovered. The first bioblitz I helped organize was at Concord, Massachusetts, in the summer of 1998, with Walden Pond as the focus. Naturalists came from all over New England. The effort was so successful and well publicized that similar events have since been conducted all over the United States (including two in New York City's Central Park) and in at least eighteen other countries.

This one took place at an elevation of around 3,700 feet on Mount Gorongosa, just below the lower fringe of rainforest. Bending to logistic necessity in this remote place—I had to get there by helicopter—we limited the time to two hours, and I served as the sole expert. I was able to identify most of the insects and spiders to their taxonomic families (such as millipedes of the family Julidae, rove beetles of the family Staphylinidae, and, of course, ants, which all belong to the family Formicidae). For some specimens I had to guess.

The event was a melee of scurrying and shouting. The children, ranging from four or five to about twelve years old, proved remarkably gifted hunters. They were eager to hear what I had to say about their discoveries. Torcida translated our talk back and forth, and at the end of the two hours I counted a total of sixty species, belonging to thirty-nine families in thirteen orders.

We found strange insects and arthropods, most very small. There were a lot of Hymenoptera (the order that includes ants, bees, and wasps), Coleoptera (beetles), and Diptera (flies). Though we saw surprisingly few ants per se, one species was identified as a rarely seen driver ant (*Dorylus bequaerti*). We also spotted a few birds, reptiles, amphibians, and one mouse.

To most of the public the word "wildlife" primarily means mammals and birds, which have suffered heavily on Mount Gorongosa. People yearn to see large wild animals, and I am no exception. But wildlife also includes the little things that run the world—the insects and other invertebrates that form the foundation of ecosystems on the land. So Gorongosa did not disappoint me. On the contrary, it fulfilled all the yearnings for adventure and discovery I have felt since my boyhood, when I was the age of my helpers on Mount Gorongosa and was venturing into the forests of Alabama and Florida with a net, spade, and collecting jars.

CARL ZIMMER

Bringing Them Back to Life

FROM *National Geographic*

ON JULY 30, 2003, a team of Spanish and French scientists reversed time. They brought an animal back from extinction, if only to watch it become extinct again. The animal they revived was a kind of wild goat known as a *bucardo,* or Pyrenean ibex. The bucardo (*Capra pyrenaica pyrenaica*) was a large, handsome creature, reaching up to 220 pounds and sporting long, gently curved horns. For thousands of years it lived high in the Pyrenees, the mountain range that divides France from Spain, where it clambered along cliffs, nibbling on leaves and stems and enduring harsh winters.

Then came the guns. Hunters drove down the bucardo population over several centuries. In 1989 Spanish scientists did a survey and concluded that there were only a dozen or so individuals left. Ten years later a single bucardo remained: a female nicknamed Celia. A team from the Ordesa and Monte Perdido National Park, led by wildlife veterinarian Alberto Fernández-Arias, caught the animal in a trap, clipped a radio collar around her neck, and released her back into the wild. Nine months later the radio collar let out a long, steady beep: the signal that Celia had died. They found her crushed beneath a fallen tree. With her death, the bucardo became officially extinct.

But Celia's cells lived on, preserved in labs in Zaragoza and Madrid. Over the next few years a team of reproductive physiologists led by José Folch injected nuclei from those cells into goat eggs emptied of their own DNA, then implanted the eggs in surrogate mothers. After fifty-seven implantations, only seven animals had become pregnant. And of those seven pregnancies, six ended in miscarriages. But one mother—a hybrid between a Spanish ibex and a goat—carried a clone of Celia to term. Folch and his colleagues performed a cesarean section and delivered the 4.5-pound clone. As Fernández-Arias held the newborn bucardo in his arms, he could see that she was struggling to take in air, her tongue jutting grotesquely out of her mouth. Despite the efforts to help her breathe, after a mere ten minutes Celia's clone died. A necropsy later revealed that one of her lungs had grown a gigantic extra lobe as solid as a piece of liver. There was nothing anyone could have done.

The dodo and the great auk, the thylacine and the Chinese river dolphin, the passenger pigeon and the imperial woodpecker —the bucardo is only one in the long list of animals humans have driven to extinction, sometimes deliberately. And with many more species now endangered, the bucardo will have much more company in the years to come. Fernández-Arias belongs to a small but passionate group of researchers who believe that cloning can help reverse that trend.

The notion of bringing vanished species back to life—some call it de-extinction—has hovered at the boundary between reality and science fiction for more than two decades, ever since the novelist Michael Crichton unleashed the dinosaurs of *Jurassic Park* on the world. For most of that time the science of de-extinction has lagged far behind the fantasy. Celia's clone is the closest that anyone has gotten to true de-extinction. Since witnessing those fleeting minutes of the clone's life, Fernández-Arias, now the head of the Aragon government's Hunting, Fishing, and Wetlands department, has been waiting for the moment when science would finally catch up, and humans might gain the ability to bring back an animal they had driven extinct.

"We are at that moment," he told me.

I met Fernández-Arias last autumn at a closed-session scientific meeting at the National Geographic Society's headquarters in Washington, DC. For the first time in history a group of geneticists, wildlife biologists, conservationists, and ethicists had gathered to discuss the possibility of de-extinction. Could it be done? Should it be done? One by one, they stood up to present remarkable advances in manipulating stem cells, in recovering ancient

DNA, in reconstructing lost genomes. As the meeting unfolded, the scientists became increasingly excited. A consensus was emerging: de-extinction is now within reach.

"It's gone very much further, very much more rapidly than anyone ever would've imagined," says Ross MacPhee, a curator of mammalogy at the American Museum of Natural History in New York. "What we really need to think about is why we would want to do this in the first place, to actually bring back a species."

In *Jurassic Park,* dinosaurs are resurrected for their entertainment value. The disastrous consequences that follow have cast a shadow over the notion of de-extinction, at least in the popular imagination. But people tend to forget that *Jurassic Park* was pure fantasy. In reality the only species we can hope to revive now are those that died within the past few tens of thousands of years and left behind remains that harbor intact cells or, at the very least, enough ancient DNA to reconstruct the creature's genome. Because of the natural rates of decay, we can never hope to retrieve the full genome of *Tyrannosaurus rex,* which vanished about 65 million years ago. The species theoretically capable of being revived all disappeared while humanity was rapidly climbing toward world domination. And especially in recent years we humans were the ones who wiped them out, by hunting them, destroying their habitats, or spreading diseases. This suggests another reason for bringing them back.

"If we're talking about species we drove extinct, then I think we have an obligation to try to do this," says Michael Archer, a paleontologist at the University of New South Wales who has championed de-extinction for years. Some people protest that reviving a species that no longer exists amounts to playing God. Archer scoffs at the notion. "I think we played God when we exterminated these animals."

Other scientists who favor de-extinction argue that there will be concrete benefits. Biological diversity is a storehouse of natural invention. Most pharmaceutical drugs, for example, were not invented from scratch—they were derived from natural compounds found in wild plant species, which are also vulnerable to extinction. Some extinct animals also performed vital services in their ecosystems, which might benefit from their return. Siberia, for example, was home 12,000 years ago to mammoths and other big grazing mammals. Back then, the landscape was not moss-dominated tundra but grassy steppes. Sergey Zimov, a Russian ecologist and director of the Northeast Science Station in Cherskiy in the Republic of Sakha, has long argued that this was no coincidence: the mammoths and numerous herbivores maintained the grassland by breaking up the soil and fertilizing it with their manure. Once they were gone, moss took over and transformed the grassland into less productive tundra.

In recent years Zimov has tried to turn back time on the tundra by bringing horses, muskoxen, and other big mammals to a region of Siberia he calls Pleistocene Park. And he would be happy to have woolly mammoths roam free there. "But only my grandchildren will see them," he says. "A mouse breeds very fast. Mammoths breed very slow. Be prepared to wait."

When Fernández-Arias first tried to bring back the bucardo ten years ago, the tools at his disposal were, in hindsight, woefully crude. It had been only seven years since the birth of Dolly the sheep, the first cloned mammal. In those early days scientists would clone an animal by taking one of its cells and inserting its DNA into an egg that had been emptied of its own genetic material. An electric shock was enough to get the egg to start dividing, after which the scientists would place the developing embryo in a surrogate mother. The vast majority of those pregnancies failed, and the few animals that were born were often beset with health problems.

Over the past decade scientists have improved their success with cloning animals, shifting the technology from high-risk science to workaday business. Researchers have also developed the ability to induce adult animal cells to return to an embryo-like state. These can be coaxed to develop into any type of cell—including eggs or sperm. The eggs can then be further manipulated to develop into full-fledged embryos.

Such technical sleights of hand make it far easier to conjure a vanished species back to life. Scientists and explorers have been talking for decades about bringing back the mammoth. Their first —and so far only—achievement was to find well-preserved mammoths in the Siberian tundra. Now, armed with the new cloning technologies, researchers at the Sooam Biotech Research Foundation in Seoul have teamed up with mammoth experts from North-Eastern Federal University in the Siberian city of Yakutsk.

Last summer they traveled up the Yana River, drilling tunnels into the frozen cliffs along the river with giant hoses. In one of those tunnels they found chunks of mammoth tissue, including bone marrow, hair, skin, and fat. The tissue is now in Seoul, where the Sooam scientists are examining it.

"If we dream about it, the ideal case would be finding a viable cell, a cell that's alive," says Sooam's Insung Hwang, who organized the Yana River expedition. If the Sooam researchers do find such a cell, they could coax it to produce millions of cells. These could be reprogrammed to grow into embryos, which could then be implanted in surrogate elephants, the mammoth's closest living relatives.

Most scientists doubt that any living cell could have survived freezing on the open tundra. But Hwang and his colleagues have a Plan B: capture an intact nucleus of a mammoth cell, which is far more likely to have been preserved than the cell itself. Cloning a mammoth from nothing but an intact nucleus, however, will be a lot trickier. The Sooam researchers will need to transfer the nucleus into an elephant egg that has had its own nucleus removed. This will require harvesting eggs from an elephant—a feat no one has yet accomplished. If the DNA inside the nucleus is well preserved enough to take control of the egg, it just might start dividing into a mammoth embryo. If the scientists can get past that hurdle, they still have the formidable task of transplanting the embryo into an elephant's womb. Then, as Zimov cautions, they will need patience. If all goes well, it will still be almost two years before they can see if the elephant will give birth to a healthy mammoth.

"The thing that I always say is, if you don't try, how would you know that it's impossible?" says Hwang.

In 1813, while traveling along the Ohio River from Hardensburgh to Louisville, John James Audubon witnessed one of the most miraculous natural phenomena of his time: a flock of passenger pigeons (*Ectopistes migratorius*) blanketing the sky. "The air was literally filled with Pigeons," he later wrote. "The light of noon-day was obscured as by an eclipse, the dung fell in spots, not unlike melting flakes of snow; and the continued buzz of wings had a tendency to lull my senses to repose."

When Audubon reached Louisville before sunset, the pigeons

were still passing overhead—and continued to do so for the next three days. "The people were all in arms," wrote Audubon. "The banks of the Ohio were crowded with men and boys, incessantly shooting at the pilgrims . . . Multitudes were thus destroyed."

In 1813 it would have been hard to imagine a species less likely to become extinct. Yet by the end of the century the red-breasted passenger pigeon was in catastrophic decline, the forests it depended upon shrinking, and its numbers dwindling from relentless hunting. In 1900 the last confirmed wild bird was shot by a boy with a BB gun. Fourteen years later, just a century and a year after Audubon marveled at their abundance, the one remaining captive passenger pigeon, a female named Martha, died at the Cincinnati Zoo.

The writer and environmentalist Stewart Brand, best known for founding the *Whole Earth Catalog* in the late 1960s, grew up in Illinois hiking in forests that just a few decades before had been aroar with the sound of the passenger pigeons' wings. "Its habitat was my habitat," he says. Two years ago Brand and his wife, Ryan Phelan, founder of the genetic testing company DNA Direct, began to wonder if it might be possible to bring the species back to life. One night over dinner with the Harvard biologist George Church, a master at manipulating DNA, they discovered that he was thinking along the same lines.

Church knew that standard cloning methods wouldn't work, since bird embryos develop inside shells, and no museum specimen of the passenger pigeon (including Martha herself, now in the Smithsonian) would likely contain a fully intact, functional genome. But he could envision a different way of re-creating the bird. Preserved specimens contain fragments of DNA. By piecing together the fragments, scientists can now read the roughly one billion letters in the passenger pigeon genome. Church can't yet synthesize an entire animal genome from scratch, but he has invented technology that allows him to make sizable chunks of DNA of any sequence he wants. He could theoretically manufacture genes for passenger pigeon traits—a gene for its long tail, for example—and splice them into the genome of a stem cell from a common rock pigeon.

Rock pigeon stem cells containing this doctored genome could be transformed into germ cells, the precursors to eggs and sperm. These could then be injected into rock pigeon eggs, where they would migrate to the developing embryos' sex organs. Squabs hatched from these eggs would look like normal rock pigeons —but they would be carrying eggs and sperm loaded with doctored DNA. When the squabs reached maturity and mated, their eggs would hatch squabs carrying unique passenger pigeon traits. These birds could then be further interbred, the scientists selecting for birds that were more and more like the vanished species.

Church's genome-retooling method could theoretically work on any species with a close living relative and a genome capable of being reconstructed. So even if the Sooam team fails to find an intact mammoth nucleus, someone might still bring the species back. Scientists already have the technology for reconstructing most of the genes it takes to make a mammoth, which could be inserted into an elephant stem cell. And there is no shortage of raw material for further experiments emerging from the Siberian permafrost. "With mammoths, it's really a dime a dozen up there," says Hendrik Poinar, an expert on mammoth DNA at McMaster University in Ontario. "It's just a matter of finances now."

Though the revival of a mammoth or a passenger pigeon is no longer mere fantasy, the reality is still years away. For another extinct species, the time frame may be much shorter. Indeed, there's at least a chance it may be back among the living before this story is published.

The animal in question is the obsession of a group of Australian scientists led by Michael Archer, who call their endeavor the Lazarus Project. Archer previously directed a highly publicized attempt to clone the thylacine, an iconic marsupial carnivore that went extinct in the 1930s. That effort managed to capture only some fragments of the thylacine's DNA. Wary of the feverish expectations that such high-profile experiments attract, Archer and his Lazarus Project collaborators kept quiet about their efforts until they had some preliminary results to offer.

That time has come. Early in January, Archer and his colleagues revealed that they were trying to revive two closely related species of Australian frog. Until their disappearance in the mid-1980s, the species shared a unique—and utterly astonishing—method of reproduction. The female frogs released a cloud of eggs, which the males fertilized, whereupon the females swallowed the eggs whole.

A hormone in the eggs triggered the female to stop making stomach acid; her stomach, in effect, became a womb. A few weeks later the female opened her mouth and regurgitated her fully formed babies. This miraculous reproductive feat gave the frogs their common names: the northern (*Rheobatrachus vitellinus*) and southern (*Rheobatrachus silus*) gastric brooding frogs.

Unfortunately, not long after researchers began to study the species, they vanished. "The frogs were there one minute, and when scientists came back, they were gone," says Andrew French, a cloning expert at the University of Melbourne and a member of the Lazarus Project.

To bring the frogs back, the project scientists are using state-of-the-art cloning methods to introduce gastric brooding frog nuclei into eggs of living Australian marsh frogs and barred frogs that have had their own genetic material removed. It's slow going, because frog eggs begin to lose their potency after just a few hours and cannot be frozen and revived. The scientists need fresh eggs, which the frogs produce only once a year, during their short breeding season.

Nevertheless, they've made progress. "Suffice it to say, we actually have embryos now of this extinct animal," says Archer. "We're pretty far down this track." The Lazarus Project scientists are confident that they just need to get more high-quality eggs to keep moving forward. "At this point it's just a numbers game," says French.

The matchless oddity of the gastric brooding frogs' reproduction drives home what we lose when a species becomes extinct. But does that mean we should bring them back? Would the world be that much richer for having female frogs that grow little frogs in their stomachs? There are tangible benefits, French argues, such as the insights the frogs might be able to provide about reproduction—insights that might someday lead to treatments for pregnant women who have trouble carrying babies to term. But for many scientists, de-extinction is a distraction from the pressing work required to stave off mass extinctions.

"There is clearly a terrible urgency to saving threatened species and habitats," says John Wiens, an evolutionary biologist at Stony Brook University in New York. "As far as I can see, there is little urgency for bringing back extinct ones. Why invest millions of dollars in bringing a handful of species back from the dead, when there are millions still waiting to be discovered, described, and protected?"

De-extinction advocates counter that the cloning and genomic engineering technologies being developed for de-extinction could also help preserve endangered species, especially ones that don't breed easily in captivity. And though cutting-edge biotechnology can be expensive when it's first developed, it has a way of becoming very cheap very fast. "Maybe some people thought polio vaccines were a distraction from iron lungs," says George Church. "It's hard in advance to say what's distraction and what's salvation."

But what would we be willing to call salvation? Even if Church and his colleagues manage to retrofit every passenger pigeon–specific trait into a rock pigeon, would the resulting creature truly be a passenger pigeon or just an engineered curiosity? If Archer and French do produce a single gastric brooding frog—if they haven't already—does that mean they've revived the species? If that frog doesn't have a mate, then it becomes an amphibian version of Celia, and its species is as good as extinct. Would it be enough to keep a population of the frogs in a lab or perhaps in a zoo, where people could gawk at it? Or would it need to be introduced back into the wild to be truly de-extinct?

"The history of putting species back after they've gone extinct in the wild is fraught with difficulty," says the conservation biologist Stuart Pimm of Duke University. A huge effort went into restoring the Arabian oryx to the wild, for example. But after the animals were returned to a refuge in central Oman in 1982, almost all were wiped out by poachers. "We had the animals, and we put them back, and the world wasn't ready," says Pimm. "Having the species solves only a tiny, tiny part of the problem."

Hunting is not the only threat that would face recovered species. For many, there's no place left to call home. The Chinese river dolphin became extinct due to pollution and other pressures from the human population on the Yangtze River. Things are just as bad there today. Around the world frogs are getting decimated by a human-spread pathogen called the chytrid fungus. If Australian biologists someday release gastric brooding frogs into their old mountain streams, they could promptly become extinct again.

"Without an environment to put re-created species back into, the whole exercise is futile and a gross waste of money," says Glenn

Albrecht, director of the Institute for Social Sustainability at Murdoch University in Australia.

Even if de-extinction proved a complete logistical success, the questions would not end. Passenger pigeons might find the rebounding forests of the eastern United States a welcoming home. But wouldn't that be, in effect, the introduction of a genetically engineered organism into the environment? Could passenger pigeons become a reservoir for a virus that might wipe out another bird species? And how would the residents of Chicago, New York, or Washington, DC, feel about a new pigeon species arriving in their cities, darkening their skies, and covering their streets with snowstorms of dung?

De-extinction advocates are pondering these questions, and most believe they need to be resolved before any major project moves forward. Hank Greely, a leading bioethicist at Stanford University, has taken a keen interest in investigating the ethical and legal implications of de-extinction. And yet for Greely, as for many others, the very fact that science has advanced to the point that such a spectacular feat is possible is a compelling reason to embrace de-extinction, not to shun it.

"What intrigues me is just that it's really cool," Greely says. "A saber-toothed cat? It would be neat to see one of those."

Contributors' Notes

Other Notable Science and Nature Writing of 2013

Contributors' Notes

Katherine Bagley is a staff reporter for InsideClimate News, covering the intersection of environmental science, politics, and policy, with an emphasis on climate change. She is also the coauthor of *Bloomberg's Hidden Legacy: Climate Change and the Future of New York City*. Her print and multimedia work has appeared in *Popular Science, Audubon, OnEarth,* and *The Scientist,* among other publications. She lives in the Hudson Valley region of New York.

Nicholas Carr is the author of *The Shallows: What the Internet Is Doing to Our Brains,* a finalist for the Pulitzer Prize, and *The Glass Cage: Automation and Us.* He has written for *The Atlantic,* the *New York Times, Wired,* and *Nature,* among other periodicals.

David Dobbs writes for *The Atlantic,* the *New York Times Magazine, National Geographic, Nature,* and other publications. He is currently writing a book with the working title *The Orchid and the Dandelion,* exploring the notion that the genes and traits underlying some of our most tormenting mood and behavior problems may also generate some of our greatest strengths and contentment. He is the author of four previous books, most recently *My Mother's Lover,* which unearths a secret affair his mother had with a doomed flight surgeon in World War II. He blogs on culture, science, and literature at Neuron Culture.

Pippa Goldschmidt, a writer based in Edinburgh, Scotland, has a PhD in astronomy. Her novel *The Falling Sky,* about an astronomer who thinks she's found evidence contradicting the Big Bang theory, was a runner-up for the Dundee International Book Prize in 2012. She has been a writer in residence at the University of Edinburgh. Her collection of short stories inspired by science will be published next year. Find out more about Goldschmidt's writing at http://www.pippagoldschmidt.co.uk.

Amy Harmon is a *New York Times* reporter who seeks to illuminate the intersection of science and society through narrative storytelling. Harmon has won two Pulitzer Prizes, one in 2008 for her series "The DNA Age," the other as part of a team for the series "How Race Is Lived in America" in 2001. Her series "Target Cancer" received the 2011 National Academies of Science award for print journalism. In 2012 her article "Autistic and Seeking a Place in an Adult World" won the Casey Medal for meritorious reporting. Harmon is the author of *Asperger Love*, which portrays a real-life relationship between two teenagers on the autism spectrum. She is the recipient of a Guggenheim Fellowship in science writing. Harmon lives in New York City with her husband and nine-year-old daughter.

Robin Marantz Henig is a contributing writer for the *New York Times Magazine*, where she has specialized in long-form science journalism, with cover stories on such topics as anxiety, death, belief in God, obesity, assisted suicide, and the science of lying. She has written nine books, most recently *Twentysomething: Why Do Young Adults Seem Stuck?*, coauthored with her daughter, Samantha Henig. Her previous books include *Pandora's Baby: How the First Test Tube Babies Sparked the Reproductive Revolution* and *The Monk in the Garden: The Lost and Found Genius of Gregor Mendel*. Henig's awards include a Guggenheim Fellowship in 2009 and a Career Achievement Award from the American Society of Journalists and Authors in 2010. She is currently serving a two-year term as president of the National Association of Science Writers.

Virginia Hughes is a journalist based in Brooklyn, New York. She is a contributing editor at *Popular Science* and *Matter* and has written for *Nature, Smithsonian,* and *Slate,* among other publications. She focuses on the brain, behavior, and genetics for her blog, *Only Human,* which is hosted by National Geographic.

Ferris Jabr is a writer based in New York City.

Sarah Stewart Johnson is an assistant professor at Georgetown University, where she studies planetary science and is currently working on a book about the exploration of Mars.

Barbara J. King is a biological anthropologist at the College of William and Mary, where she teaches half-time in order to devote more time to freelance science writing. Her research, initially focused on wild baboons and captive African apes, now includes cognition and emotion in animals from chickens to chimpanzees. Her latest book, *How Animals Grieve*, has been featured on BBC-TV and in other international media. King contributes weekly to NPR.org's 13.7 Cosmos & Culture blog and writes regularly for *The Times Literary Supplement*. At home in Virginia, she and her husband care for rescued cats.

Barbara Kingsolver's fourteen books of fiction, poetry, and creative nonfiction include the novels *The Bean Trees, The Poisonwood Bible, The Lacuna*, and her most recent novel, *Flight Behavior*. Translated into more than twenty languages, her work has won a devoted worldwide readership, a place in the core English literature curriculum, and many awards, including the National Humanities Medal. Her fiction has been three times shortlisted for and once a winner of Britain's Orange Prize.

Maggie Koerth-Baker is a freelance science journalist. She writes the monthly column "Eureka" for the *New York Times Magazine* and is also the science editor at BoingBoing.net, a technology and culture blog with 6 million monthly readers. In 2012 she published *Before the Lights Go Out*, a book about the future of energy and the United States electric grid.

Elizabeth Kolbert is a staff writer for *The New Yorker* and the author of *The Sixth Extinction* and *Field Notes from a Catastrophe: Man, Nature, and Climate Change*. She is a two-time National Magazine Award winner and has received a Heinz Award, a Guggenheim Fellowship, and a Lannan Literary Fellowship. Kolbert lives in Williamstown, Massachusetts.

Joshua Lang is a medical student at the University of California, San Francisco. He has lived in barns on horse-racing tracks in Chicago, chronicled a Sudanese woman's struggle to find health care in Alaska, slept in underground laboratories in Wisconsin, and trained at an abortion clinic in New Mexico. His writing on social and scientific topics related to medicine has appeared in publications including *The Atlantic* and the *New York Times Magazine*. His academic research focuses on early identification and prevention of chronic diseases in persons living with HIV.

Maryn McKenna is an independent journalist specializing in public health, global health, and food policy. She is a contributing writer for *Wired* and for *National Geographic*'s food-writing platform *The Plate*, and she writes for *Scientific American, Nature, Slate*, the *Guardian, The Atlantic*, and other publications in the United States and Europe. She is the author of the award-winning books *Superbug,* about the global rise of antibiotic resistance, and *Beating Back the Devil,* about the U.S. Centers for Disease Control and Prevention, and is currently working on a book about food production. She is a senior fellow of the Schuster Institute for Investigative Journalism at Brandeis University and has been a research fellow at MIT and the University of Michigan.

Seth Mnookin is the associate director of the MIT Graduate Program in Science Writing. His most recent book, *The Panic Virus: The True Story of the Vaccine-Autism Controversy,* won the National Association of Science Writers' Science in Society book award in 2012. He is also the author of the 2006 national bestseller *Feeding the Monster,* about the Boston Red Sox, and *Hard News,* about the *New York Times.* He lives with his wife, their two children, and their nine-year-old adopted pit bull in Brookline, Massachusetts.

"Each life is an encyclopedia," wrote Italo Calvino. "A library, an inventory of objects, a series of styles." With such inspiration stuffed deep in his pouch, **Justin Nobel** has begun a project he calls The Decalogy, a set of ten books on topics close to the heart. He's presently at work on the second book, a string of vignettes from forgotten small towns of the American South, as well as the third, a collection of tales about the weather. Nobel lives in New Orleans with MissKarret and Jazzy-B, Fern, Ishtar, and Lady Sal.

Fred Pearce is a longtime journalist and author on issues of the environment and development. He is an environment consultant at *New Scientist* and a regular contributor to *Yale e360.* His books include *The Coming Population Crash, When the Rivers Run Dry,* and *The Landgrabbers.*

Corey S. Powell is editor at large at *Discover* magazine, where he writes the "Out There" column and blog. He is also the acting editor of *American Scientist* and a visiting scholar at NYU's Science Health and Environmental Reporting Program. His writing appears in *Smithsonian, Popular Science,* and *Slate* in addition to *Nautilus.* He is the author of *God in the Equation,* an exploration of the spiritual impulse in modern cosmology.

Roy Scranton's writing has appeared in the *New York Times, Boston Review, Sierra Magazine, Prairie Schooner, Theory & Event, Kenyon Review, The Appendix,* and other periodicals. He edited and contributed to *Fire and Forget: Short Stories from the Long War,* an anthology of literary fiction by veterans of the Iraq and Afghanistan wars, and is currently working on a book-length version of "Learning How to Die in the Anthropocene," to be published in 2015. Scranton lives in New York and is finishing a PhD in English at Princeton.

Kate Sheppard is a senior reporter and the environment and energy editor at the *Huffington Post*. She previously reported for *Mother Jones, Grist,* and the *American Prospect*. Her reporting has been recognized with awards from the Society of Environmental Journalists (SEJ), the Online News Association, and Planned Parenthood, and she is a board member of SEJ. She was raised on a vegetable farm in New Jersey but now calls Washington, DC, home.

Born in Bridgeport, Connecticut, nature writer **Bill Sherwonit** has called Alaska home since 1982. He has contributed essays and articles to a wide variety of newspapers, magazines, journals, and anthologies and is the author of more than a dozen books. His most recent books include *Living with Wildness: An Alaskan Odyssey* and *Changing Paths: Travels and Meditations in Alaska's Arctic Wilderness*. A collection of his essays, *Animal Stories: Encounters with Alaska's Wildlife* will be published in fall 2014. See www.billsherwonit.alaskawriters.com.

Rebecca Solnit is the author of fifteen books about environment, landscape, community, art, politics, hope, and memory, including *The Faraway Nearby; A Paradise Built in Hell: The Extraordinary Communities That Arise in Disaster; A Field Guide to Getting Lost; Wanderlust: A History of Walking;* and *River of Shadows: Eadweard Muybridge and the Technological Wild West* (for which she received a Guggenheim Fellowship, the National Book Critics Circle Award in criticism, and a Lannan Literary Award). A product of the California public education system from kindergarten to graduate school, she is a contributing editor to *Harper's Magazine* and a frequent contributor to the political site Tomdispatch.com.

David Treuer is an Ojibwe from Leech Lake Reservation in northern Minnesota. He is the author of three novels, a book of criticism, and, most recently, a major work of nonfiction entitled *Rez Life: An Indian's Journey Through Reservation Life*. His work has appeared in *Harper's Magazine, Esquire, Orion,* the *Washington Post,* the *New York Times,* and the *Los Angeles Times.*

Edward Osborne Wilson is generally recognized as one of the leading biologists in the world. He is acknowledged as the creator of two scientific disciplines (island biogeography and sociobiology), three unifying concepts for science and the humanities jointly (biophilia, biodiversity studies, and consilience), and one major technological advance in the study of global biodiversity (the Encyclopedia of Life). Among more than one hundred awards he has received worldwide are the U.S. National Medal of Science, the Crafoord Prize (the equivalent of the Nobel Prize, for ecology) of the Royal Swedish Academy of Sciences, and the International Prize of Biology of Japan; in letters, he has received two Pulitzer prizes in nonfiction, the Nonino and Serono prizes of Italy, and the International Cosmos Prize of Japan. He is currently Honorary Curator in Entomology and University Research Professor Emeritus, Harvard University.

Carl Zimmer is a columnist for the *New York Times* and writes features and a blog for *National Geographic.* He is the author of twelve books, including *Parasite Rex* and *Evolution: Making Sense of Life,* a textbook he coauthored with the biologist Doug Emlen. He has won the American Association for the Advancement of Science Journalism Award three times and has also won the National Academies Communication Award. A lecturer at Yale University, Zimmer lives in Guilford, Connecticut, with his wife, Grace Zimmer, and their two daughters.

Other Notable Science and Nature Writing of 2013

SELECTED BY TIM FOLGER

AMANDA MASCARELLI
  Growing Up with Pesticides. *Science.* August 16
BILL MCKIBBEN
  A Moral Atmosphere. *Orion.* March/April
KENNETH MILLER
  Mushroom Manifesto. *Discover.* July/August
JOHN MOIR
  Nature's Blinded Visionaries. *Catamaran.* Spring
MITCH MOXLEY
  The Rat Hunters of New York. *Roads & Kingdoms.* 2013

NICK NEELY
  The Edge Effect. *Missouri Review.* Winter
WENDEE NICOLE
  Game On! *Ensia.* March 26
MICHELLE NIJHUIS
  The Ghost Commune. *Aeon Magazine.* October 31
  Swimming in Sperm and Eggs. *Slate.* February 26

CAITLIN O'CONNELL-RODWELL
  Mean Girls. *Smithsonian.* March
DENNIS OVERBYE
  A Quantum of Solace. *New York Times.* July 1

KHARUNYA PARAMAGURU
  The Battle over Global Warming Is All in Your Head. *Time.* August 19
COREY S. POWELL
  The Sculpture on the Moon. *Slate.* December 16

DAVID QUAMMEN
  The Wild Life of a Bonobo. *National Geographic.* March

BENJAMIN RACHLIN
  The Accidental Beekeeper. *Virginia Quarterly Review.* Summer
MARY ROACH
  The Marvels in Your Mouth. *New York Times.* March 25
LESLIE ROBERTS
  The Art of Eradicating Polio. *Science.* October 4
JULIAN RUBINSTEIN
  Operation Easter. *The New Yorker,* July 22.

CAMERON M. SMITH
  Starship Humanity. *Scientific American.* January
DON STAP
  Site Fidelity. *Fourth Genre.* Fall
MANIL SURI
  How to Fall in Love with Math. *New York Times.* September 15

THE BEST AMERICAN SERIES®

FIRST, BEST, AND BEST-SELLING

The Best American series is the premier annual showcase for the country's finest short fiction and nonfiction. Each volume's series editor selects notable works from hundreds of magazines, journals, and websites. A special guest editor, a leading writer in the field, then chooses the best twenty or so pieces to publish. This unique system has made the Best American series the most respected—and most popular—of its kind.

Look for these best-selling titles in the Best American series:

The Best American Comics

The Best American Essays

The Best American Infographics

The Best American Mystery Stories

The Best American Nonrequired Reading

The Best American Science and Nature Writing

The Best American Short Stories

The Best American Sports Writing

The Best American Travel Writing

Available in print and e-book wherever books are sold.
Visit our website: *www.hmhbooks.com/hmh/site/bas*